This book belongs to:

_____

Table of Contents

Number Sense
Number Sense Checklist 4
Math Is Everywhere! 5
Number Word Search 6
Reading and Writing Number Words 7
Using Ten Frames to Count to 10 8
Using Ten Frames to Count to 20 10
Tens and Ones to 50 12
Counting Forward to 100 14
Counting Backward from 100 15
Counting by 2s to 100 16
Counting by 5s to 100 17
Ordering Numbers 18
Comparing Numbers from 1 to 10 20
Comparing Numbers to 50 21
Estimating and Counting 22
Show What You Know! 24

Fair-Sharing and Fractions
Fair-Sharing and Fractions Checklist 26
Exploring Fair Sharing 27
More Fair Sharing 30
Exploring More Fair Sharing 32
Exploring Equal Parts: Halves 34
Exploring Equal Parts: Fourths 36
Exploring Equal Parts: Halves and Fourths 38
Colour the Fractions 40
Working with Fractions 41
Ordering Fractions 42
Comparing Fractions 43
Show What You Know! 44

Operations
Operations Checklist 46
Plus Zero Addition Strategy 47
Count On Addition Strategy 48
Using a Number Line to Add 49
Turn Around Addition Strategy 50
Counting Doubles Addition Strategy 51
Doubles Plus One Addition Strategy 52
Draw a Picture Addition Strategy 53
Addition Word Problems 54
Adding Ten More 57
Addition: Sums to 10 58
Addition: Sums to 20 59
Minus Zero Subtraction Strategy 60
Counting Back Subtraction Strategy 61
Using a Number Line to Subtract 62
A Number Minus Itself 63
Doubles Subtraction Strategy 64
Draw a Picture Subtraction Strategy 65
Subtraction Word Problems 66
Ten More, Ten Less 69
Subtracting Numbers from 0 to 10 70
Subtracting Numbers from 0 to 20 71
Show What You Know — Sums Tests 72
Show What You Know — Subtraction Tests .. 73
Exploring Equal Groups 74

Patterns and Relationships
Patterns and Relationships Checklist 78
Patterns Are All Around Us! 79
Patterns in Everyday Life 81
Extend the Pattern 82
Growing Patterns 84
Shrinking Patterns 85
How Does the Pattern Attribute Change? 86
Describing the Pattern Rule 88
Identifying and Describing Patterns 89
Complete the Pattern 90
Create a Pattern 92
Create a Pattern Rule 93
Skip Counting to 100 94
Line Patterns 95
More and Fewer Objects 96
Show What You Know! 98

Variables, Equalities, and Inequalities
Variables, Equalities, and Inequalities Checklist 100
Understanding Things That Stay the Same and
    Things That Can Change 101
Balance It! 102
Making Addition Sentences 104
Making Subtraction Sentences 106
Number Fact Families 108
Is It Equal? 110
Show What You Know! 112

Coding Skills
Coding Skills Checklist 114

Table of Contents

Coding in Everyday Life 115
Read the Code 116
Exploring Writing Code 118
Alter the Code 120
Show What You Know! 122

Money Concepts
Money Concepts Checklist 124
Canadian Coins 125
Matching Canadian Coins 127
Counting Nickels 128
Counting Dimes 129
Comparing Money 130
Comparing Coin Sizes 131
Canadian Bills 132
Comparing Bill Values 134
Order the Coins and Bills 135
Show What You Know! 136

Data Literacy
Data Literacy Checklist 138
Create Your Own Survey! 139
Exploring Pictographs 140
Exploring Tally Charts 142
Exploring Bar Graphs 144
Collecting Data 146
Sorting Objects 148
Ordering Objects 151
Show What You Know! 152

Probability
Probability Checklist 154
Thinking About Likelihood 155
Exploring Probability 156
Show What You Know! 158

Geometric and Spatial Reasoning
Geometric and Spatial Reasoning Checklist ... 160
Colouring 2D Shapes 161
Exploring 2D Shapes 162
Shapes All Around Us! 166
Take a 2D Shape Walk 167
Sorting 2D Shapes 168
Draw the Other Half 169
Exploring 3D Figures 170

Sorting 3D Figures 171
Where Is the Hamster? 172
Where Is the Dog? 173
Positional Words: Follow the Instructions 174
Exploring Location and Movement 175
Show What You Know! 176

Measurement
Measurement Checklist 178
Exploring Measurement: Non-Standard Units 179
Exploring Measurement: Centimetres 181
Comparing Length 183
Comparing Height 184
Comparing Capacity 185
Comparing Mass 186
Show What You Know! 188

Time
Time Checklist 190
Complete a Clock Face 191
Telling Time to the Hour 192
Telling Time to the Half Hour 194
Exploring Digital Clocks 196
A.M. and P.M. 197
Months of the Year 198
Reading a Calendar 199
Show What You Know! 200

Award 202
Answer Pages 203

I Can Checklist:
Number Sense

Read and draw numbers all the way up to 50. For example:
- *I can* read numbers up to 50 like 1, 10, 25, or 50.
- *I can* show numbers up to 50 using objects, pictures, or drawings.
- *I can* talk about how numbers are used in everyday life, like counting money.

Compose and decompose whole numbers up to and including 50. For example:
- *I can* break down a number like 25 into smaller parts, like 20 and 5.
- *I can* put smaller parts together to make a number, like 20 and 5 make 25.
- *I can* use tools like blocks, pictures, or my fingers, to understand how numbers work.

Look at numbers up to 50 and figure out which ones are bigger, smaller, or the same, in different situations. For example:
- *I can* tell if a number is bigger or smaller than another number.
- *I can* put numbers in order from smallest to biggest or biggest to smallest.
- *I can* use numbers to compare things, like saying I have more or fewer toys.

Make a smart guess about how many things are in a group of up to 50 things, then count to see if your guess was close. For example:
- *I can* make a guess about how many things there are in a group, like guessing there are about 20 candies in a jar.
- *I can* count the objects to see if my guess was close or not.
- *I can* use counting to check if my estimate was right or wrong.

Count up to 50 by 1s, 2s, 5s, and 10s using different ways and tools. For example:
- *I can* count from 1 to 50 by saying one number after another, like 1, 2, 3, and so on.
- *I can* skip count by 2s, like 2, 4, 6, 8, and so on.
- *I can* skip count by 5s, like 5, 10, 15, 20, and so on.
- *I can* skip count by 10s, like 10, 20, 30, 40, and so on.
- *I can* use different tools, like number lines or my fingers, to help me count.

Math Is Everywhere!

Find pictures in magazines that show how we use numbers every day. Cut them out and stick them on paper.

I can describe how numbers are used in everyday life.

Number Word Search

Find and circle the number words below.

T	S	Y	F	I	V	E	S
H	E	O	N	E	I	H	Q
R	V	W	S	R	W	G	X
E	E	S	N	I	N	E	F
E	N	Z	E	R	O	W	O
J	K	E	I	G	H	T	U
T	W	O	L	C	B	A	R
O	S	I	X	P	T	E	N

Cross the numbers off the list as you find them.

zero   one   two   three   four
five   six   seven   eight   nine   ten

Reading and Writing Number Words

Print the number for the number word.
Answer the questions using the number words from the box.

zero	one	two	
			three
four	five	six	
			seven
eight	nine	ten	

1 How old are you?
_____
_____

2 How many pets do you have?
_____
_____

3 How many people are there in your family?
_____
_____

4 How many days are there in one week?
_____
_____

5 How many fingers do you have altogether?
_____
_____

I can read and write number words.

Using Ten Frames to Count to 10

1 Trace the number.
Draw ●s in the ten frame to show the number.

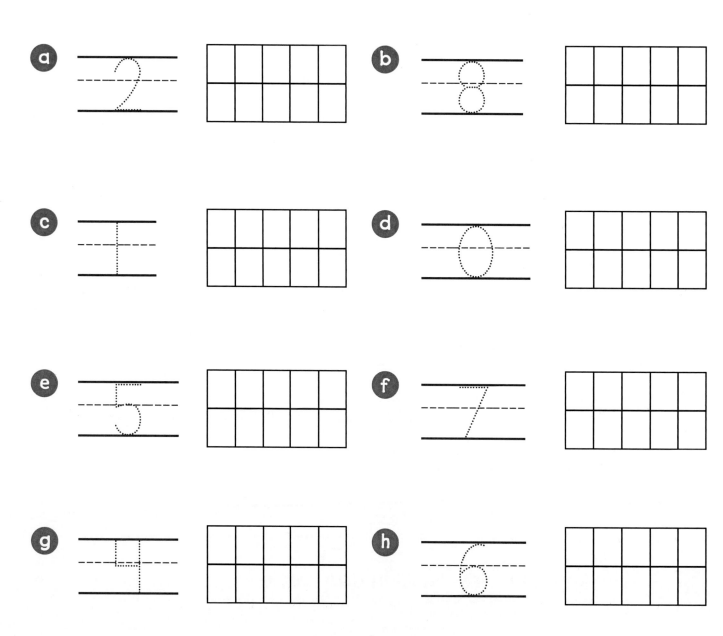

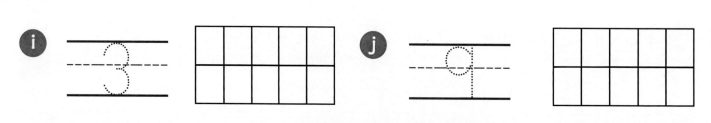

Using Ten Frames to Count to 10 (continued)

2 How many counters are in the ten frame? Print the number on the line.

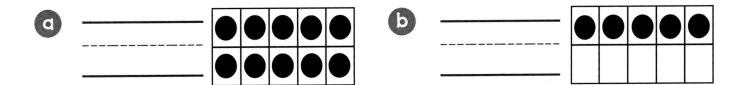

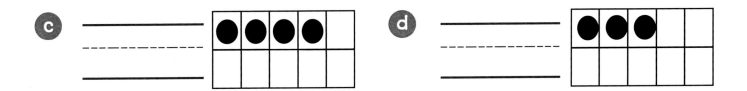

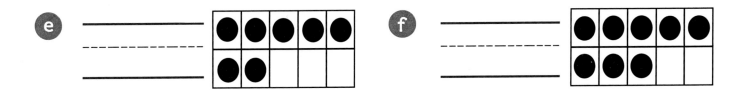

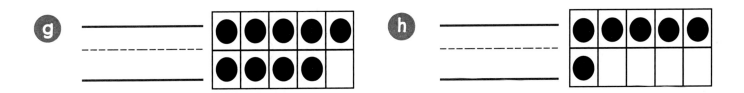

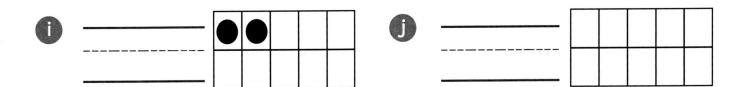

I can count to 10 using ten frames.

Using Ten Frames to Count to 20

1 How many counters are in the ten frames?
Print the number on the line.

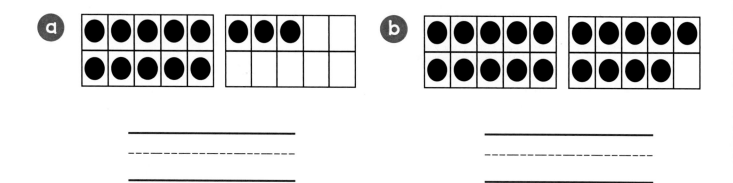

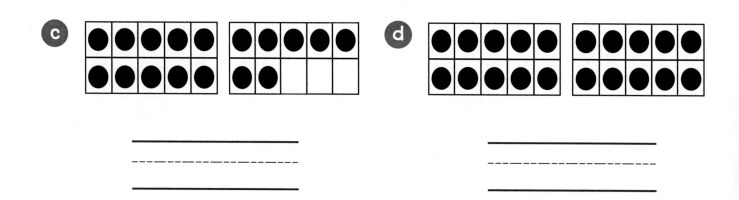

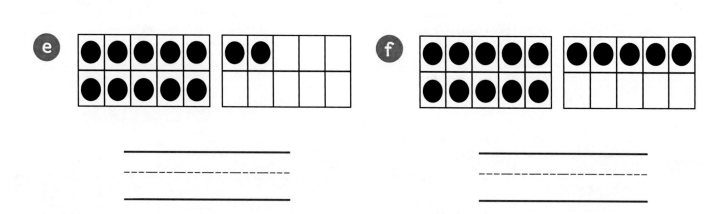

Using Ten Frames to Count to 20 (continued)

2 Draw ●s in the ten frames to show each number.

a) 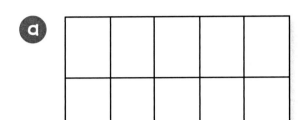 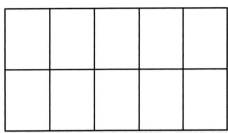  16

b) 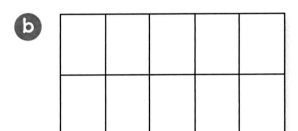 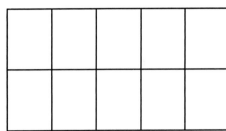  14

c) 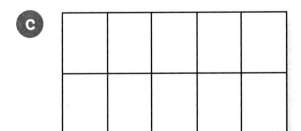 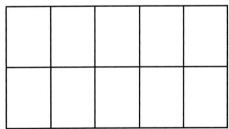  11

d) 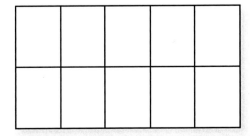 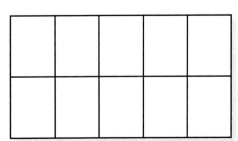 20

I can use ten frames to count to 20.

Tens and Ones to 50

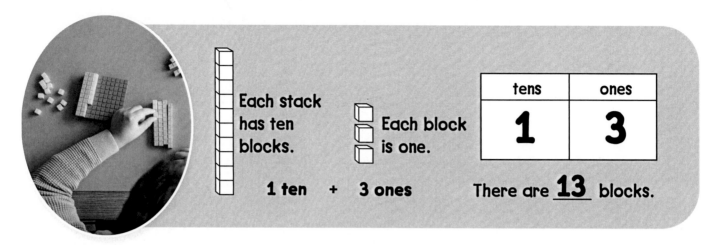

1 ten + 3 ones

tens	ones
1	3

There are __13__ blocks.

1 Count the tens and ones. Write how many blocks in all.

a

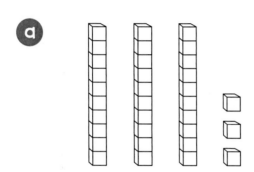

tens	ones

There are _____ blocks.

b

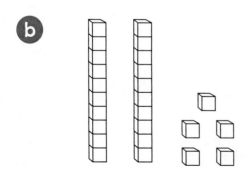

tens	ones

There are _____ blocks.

I can count tens and ones to 50.

Tens and Ones to 50 (continued)

2 Count the tens and ones. Write how many blocks in all.

a

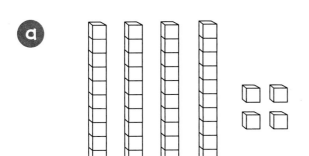

tens	ones

There are _____ blocks.

b

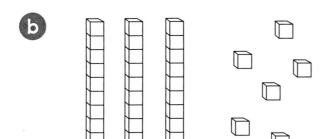

tens	ones

There are _____ blocks.

c

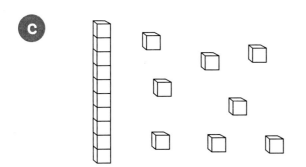

tens	ones

There are _____ blocks.

I can count tens and ones to 50.

Counting Forward to 100

1 Count forward by **1s**. Fill in the missing numbers.

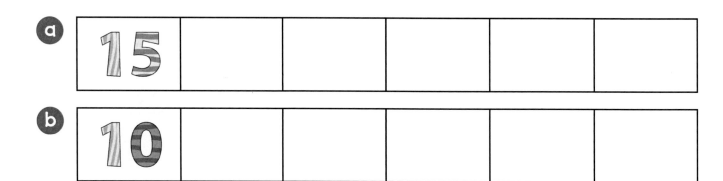

2 Count forward by **5s**. Fill in the missing numbers.

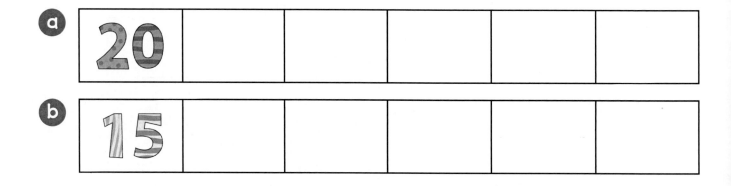

3 Count forward by **10s**. Fill in the missing numbers.

a) 10 □ □ □ □ □

b) 50 □ □ □ □ □

Counting Backward from 100

1 Count backward by **1s**. Fill in the missing numbers.

a) | 8 | | | | | |

b) | 10 | | | | | |

2 Count backward by **5s**. Fill in the missing numbers.

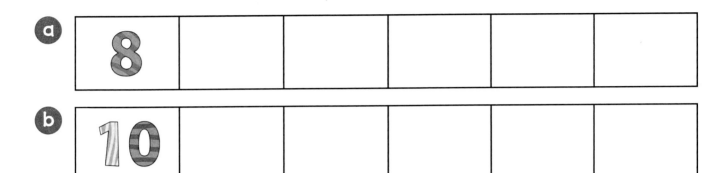

a) | 50 | | | | | |

b) | 30 | | | | | |

3 Count backward by **10s**. Fill in the missing numbers.

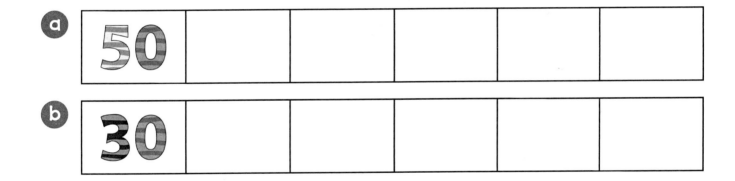

a) | 60 | | | | | |

b) | 80 | | | | | |

I can count backward from 100.

Counting by 2s to 100

1 Count out loud by **2s** to 100. Connect the dots.

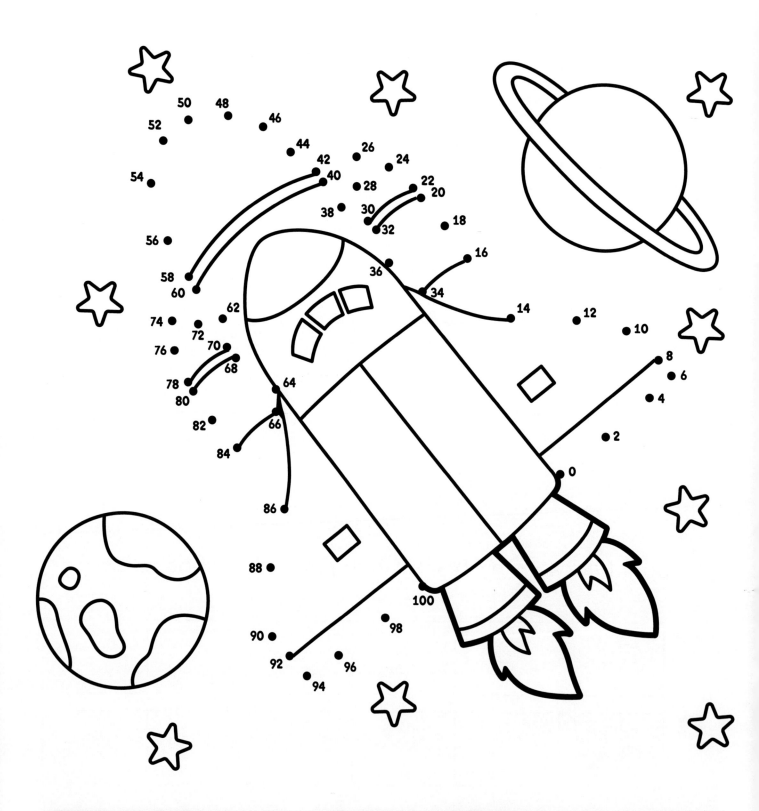

I can count by 2s to 100.

Counting by 5s to 100

1 Count out loud by **5s** to 100. Connect the dots.

I can count by 5s to 100.

Ordering Numbers

1 Fill in the missing numbers from the chart.

1	2	3		5	6	7	8	9	10
11	12		14	15	16	17	18		20
	22	23	24	25	26		28	29	30
31		33	34	35	36	37		39	40
41	42	43	44	45		47	48	49	

2 Fill in the blank below.

a Just before

____ , 11 , 12

b Just after

48 , 49 , ____

c Between

20 , ____ , 22

I can order numbers.

Ordering Numbers (continued)

3 Order the numbers from **least** to **greatest**.

a) 50 34 40 ____ ____ ____

b) 20 48 22 ____ ____ ____

c) 17 14 21 ____ ____ ____

4 Order the numbers from **greatest** to **least**.

a) 1 49 34 ____ ____ ____

b) 33 52 27 ____ ____ ____

c) 17 20 36 ____ ____ ____

I can order numbers.

Comparing Numbers from 1 to 10

Compare the numbers by counting.

3 is **less than** 5   4 is **greater than** 2   6 is **equal to** 6

1 Compare the numbers. Print the correct symbol in the box.

| greater than | less than | equal to |

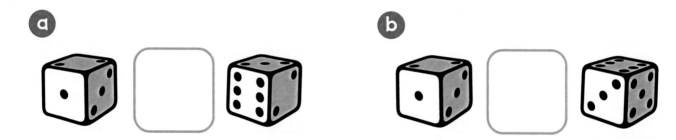

I can compare numbers from 1 to 10.

Comparing Numbers to 50

1 Compare the numbers. Print the correct symbol in the box.

greater than   less than   equal to

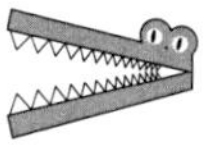

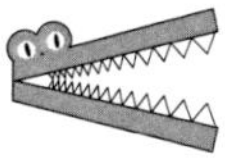

a) 43 ☐ 10

b) 19 ☐ 28

c)

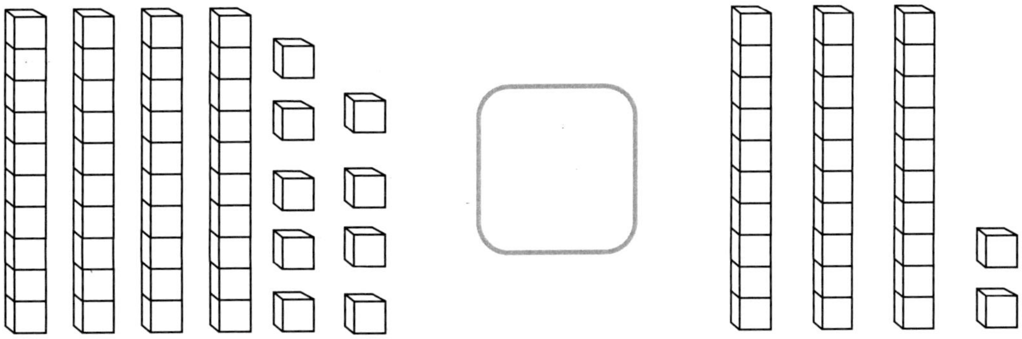

d)

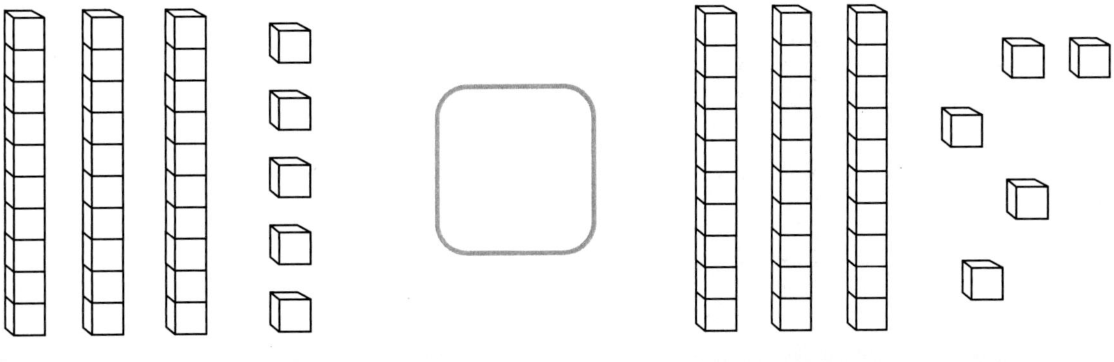

I can compare numbers up to 50.

Estimating and Counting

Estimation is a smart guess. It is like thinking, "about how many?" without counting all the way.

1 Estimate, then count the total number of objects in the set.

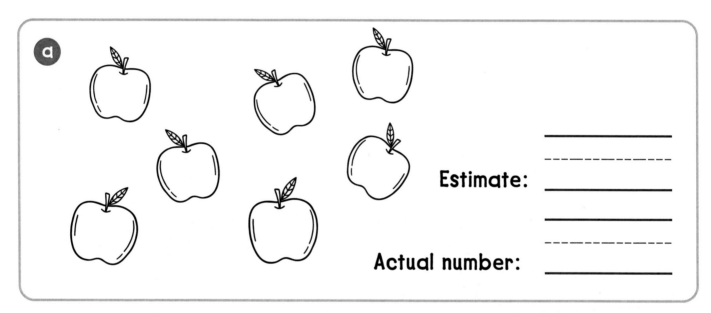

a

Estimate: _____

Actual number: _____

b

Estimate: _____

Actual number: _____

I can estimate and count.

Estimating and Counting (continued)

2 Estimate, then count the total number of objects in each set.

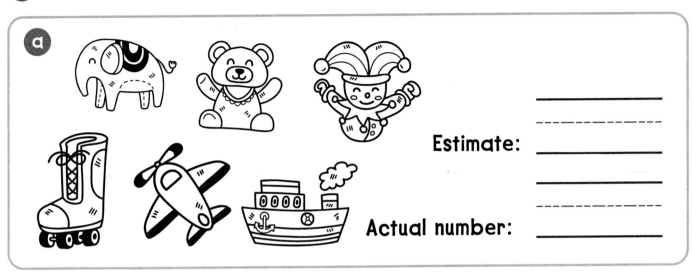

Estimate: _____

Actual number: _____

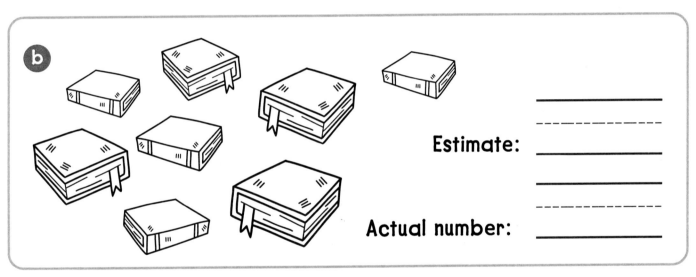

Estimate: _____

Actual number: _____

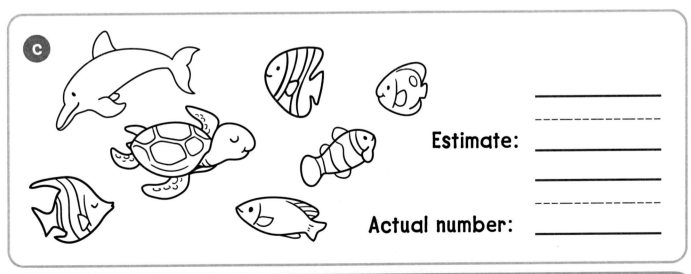

Estimate: _____

Actual number: _____

I can estimate and count.

Show What You Know!

1 How many counters are in the ten frames?
Print the number on the line.

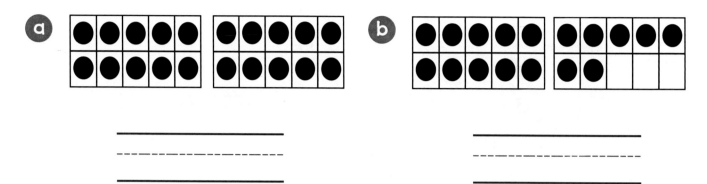

2 Count the tens and ones. Write how many blocks in all.

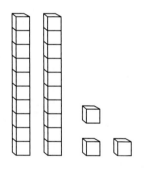

tens	ones

There are _____ blocks.

3 Estimate, then count the total number of objects in the set.

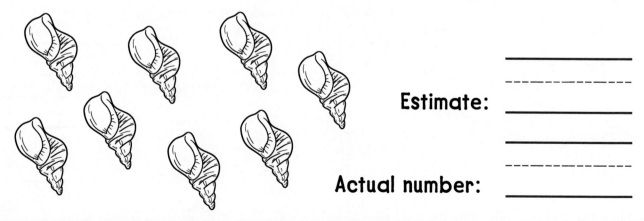

Estimate: _____

Actual number: _____

Show What You Know! (continued)

4 Compare the numbers. Print the correct symbol in the box.

| greater than | less than | equal to |

a

b

5 Order the numbers from **greatest** to **least**.

a 50   22   40   ___ ___ ___

b 48   49   52   ___ ___ ___

6 Write the word for the number 10.   _____

I Can Checklist:
Fair Sharing and Fractions

Draw pictures to help figure out how to share things between 2 or 4 friends, when there are 1 or 2 pieces left over. For example:

- *I can* draw pictures to show how things can be shared equally between 2 people, like sharing 6 cookies so each person gets 3 cookies and there is 1 left over.
- *I can* draw pictures to show how things can be shared equally between 4 people, like sharing 10 toys so each person gets 2 toys and there are 2 left over.

Understand that half of something is the same as two quarters of that same thing when you are sharing fairly. For example:

- *I can* understand that when we split something into 2 equal parts, each part is called one half.
- *I can* understand that when we split something into 4 equal parts, each part is called one fourth or one quarter. Two fourths is the same as one half.

Draw pictures to help you understand and sort small pieces of a whole thing. Imagine you are sharing it with different numbers of friends, up to 10 friends. For example:

- *I can* draw pictures to show how a whole can be divided into equal parts, like dividing a pizza into 4 slices or dividing a chocolate bar into 8 pieces.
- *I can* compare unit fractions by looking at the size of the parts, like knowing that one half is bigger than one fourth.
- *I can* order unit fractions from smallest to largest or largest to smallest, like knowing that one eighth is smaller than one fourth.

Exploring Fair Sharing

Fair sharing in math means making sure that everyone gets the same amount of something when it is divided.

1 Use four different colours to show how many cupcakes each friend should get to make it a fair share.

I can identify fair sharing.

Exploring Fair Sharing (continued)

2 Share the food equally.
Write how many each child will get.

I can identify fair sharing.

Exploring Fair Sharing (continued)

3 Share the books equally.
Write how many each child will get.

I can identify fair sharing.

More Fair Sharing

1. Are the friends sharing the items fairly? Circle **YES** or **NO**.

a

YES   NO

b

YES   NO

I can identify fair sharing.

More Fair Sharing (continued)

2 Are the friends sharing the items fairly? Circle **YES** or **NO**.

a) YES   NO

b) YES   NO

I can identify fair sharing.

Exploring More Fair Sharing

1 Circle the equal share each child will get.
Write how many are left over.

a There are **4 children.**

🪣s left over: _____

b There are **2 children.**

🦢s left over: _____

c There are **4 children.**

⚽s left over: _____

I can identify fair sharing with left over objects.

Exploring More Fair Sharing (continued)

2 Circle the equal share each child will get. Write how many are left over.

a There are **2** children.

🍦s left over: _____

b There are **2** children.

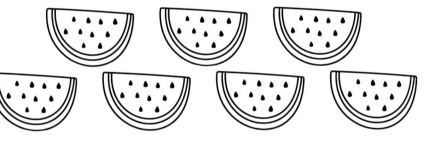

🍉s left over: _____

c There are **4** children.

👓s left over: _____

I can identify fair sharing with left over objects.

Exploring Equal Parts: Halves

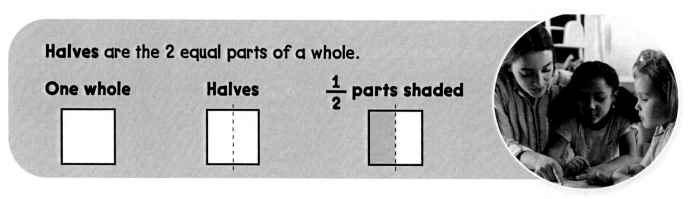

Halves are the 2 equal parts of a whole.

One whole   Halves $\frac{1}{2}$ parts shaded

1 Draw a line to show two equal parts on each of the shapes.

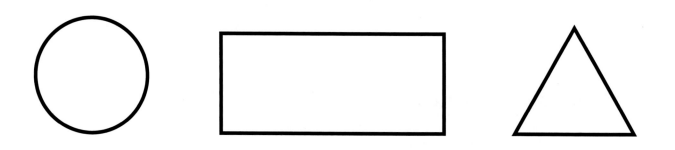

2 Draw lines to show two equal parts two different ways.

a   b

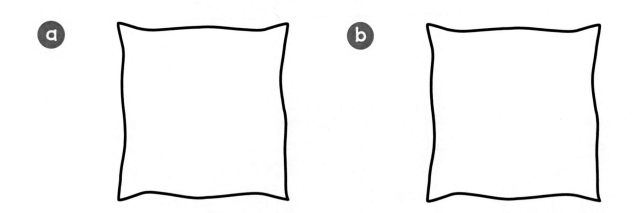

I can identify equal halves.

Exploring Equal Parts: Halves (continued)

3 Circle and colour one half of each set.

a

Half of 8: _____

b

Half of 4: _____

c

Half of 10: _____

I can identify equal halves.

Exploring Equal Parts: Fourths

Fourths are the 4 equal parts of a whole.

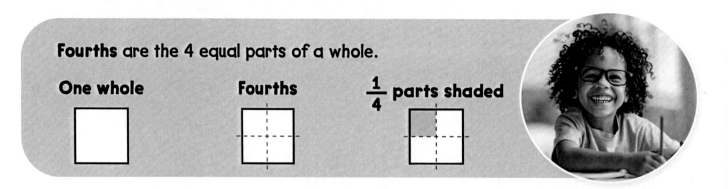

1 Draw two lines to show four equal parts on each of the shapes.

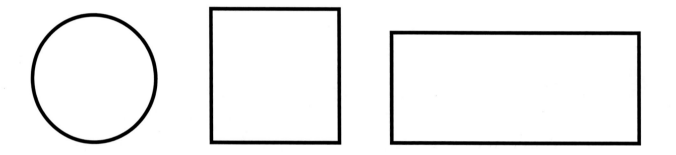

2 Draw lines to show four equal parts two different ways for the shape.

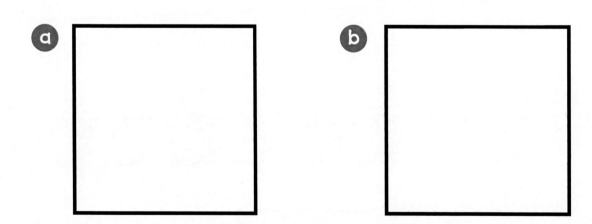

I can explore equal fourths.

Exploring Equal Parts: Fourths (continued)

3 Circle and colour one fourth of each set.

a

One fourth of 4: _____

b

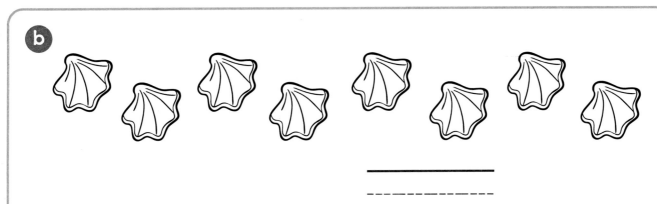

One fourth of 8: _____

c

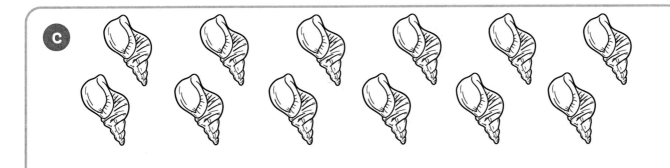

One fourth of 12: _____

Exploring Equal Parts:
Halves and Fourths

1 Colour the halves 🖍 red. Colour the fourths 🖍 blue.

I can explore equal halves and fourths.

Exploring Equal Parts:
Halves and Fourths (continued)

2. Colour $\frac{1}{2}$ red.   Colour $\frac{2}{4}$ blue.

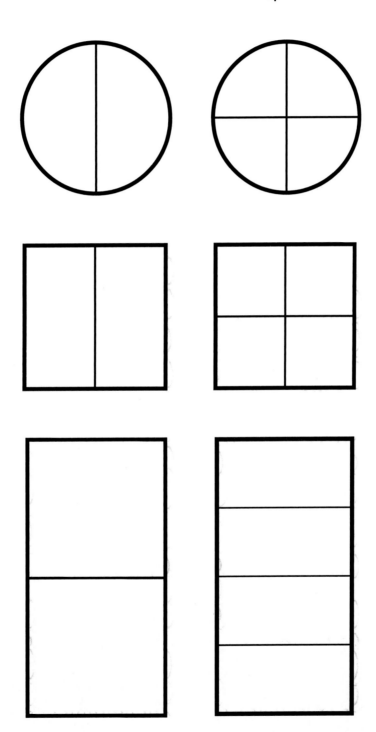

I can explore equal halves and fourths.

Colour the Fractions

$\frac{1}{2}$ means one out of two parts, or one half is shaded.

Colour in the shape to match the fraction shown.

1 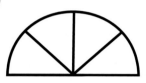 $\frac{2}{4}$

2 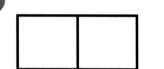 $\frac{1}{2}$

3 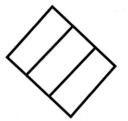 $\frac{1}{3}$

4  $\frac{2}{3}$

5 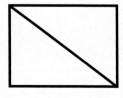 $\frac{1}{2}$

6 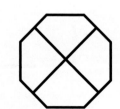 $\frac{3}{4}$

I can identify fractions.

Working with Fractions

1 Circle the shape that shows **two equal** parts.

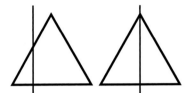

2 Circle the shape that shows **equal** parts.

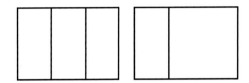

3 What part of the shape is **shaded**? Circle the fraction.

$$\frac{1}{2} \quad \frac{1}{3} \quad \frac{1}{4}$$

4 What part of the shape is **shaded**? Circle the fraction.

$$\frac{1}{2} \quad \frac{1}{3} \quad \frac{1}{4}$$

5 Divide the shape into **two equal** parts.

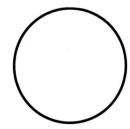

6 Circle the shape that shows **two equal** parts.

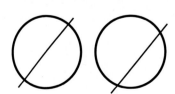

I can identify fractions.

Ordering Fractions

Look at the shaded fraction of the shape. Number the fractions from **smallest** to **biggest**. Write 1, 2, 3, and so on.

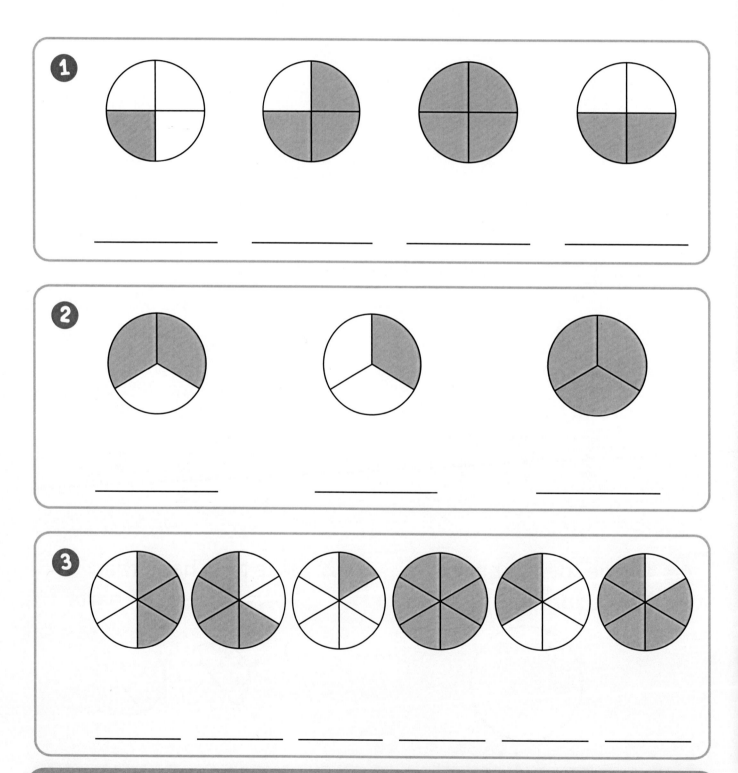

Comparing Fractions

Compare the fractions. Print the correct symbol in the box.

greater than   less than   equal to

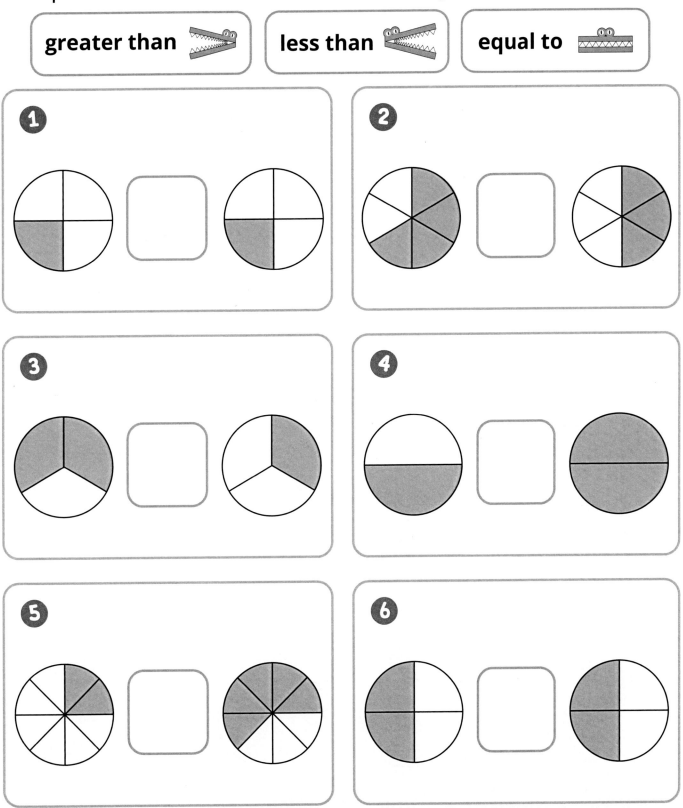

Show What You Know!

1 Colour in the shape to match the fraction shown.

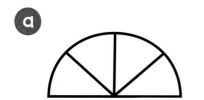

2 Compare the fractions. Print the correct symbol in the box.

 greater than   less than   equal to

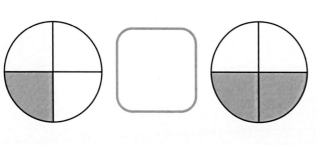

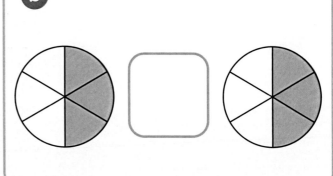

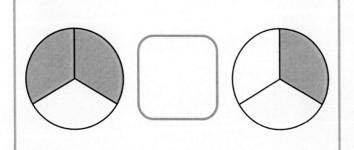

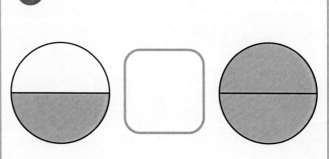

Show What You Know! (continued)

3 Are the friends sharing the items fairly? Circle **YES** or **NO**.

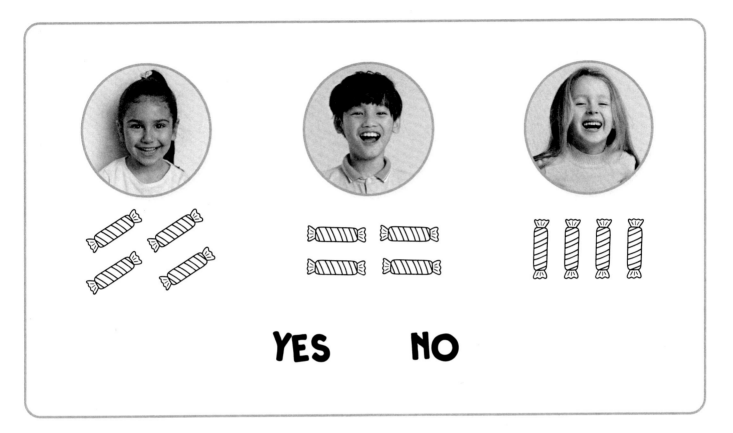

YES   NO

4 Look at the shaded fraction of the shape. Number the fractions from smallest to biggest. Write 1, 2, 3, and so on.

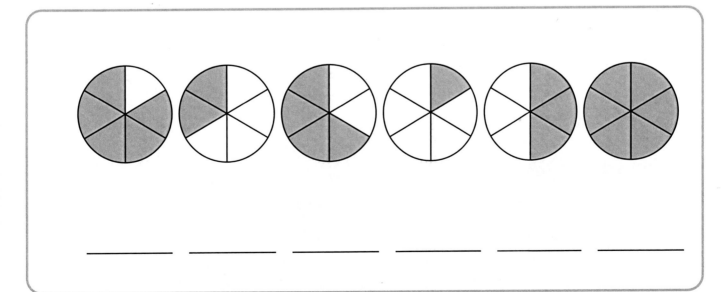

I Can Checklist:
Operations

Properties and Relationships

I can use adding and subtracting to solve problems and check my answers.

- ***For example***, *if I have 3 apples and my friend gives me 2 more, I can use addition to find out that I have 5 apples now. If I want to know how many apples my friend has, I can use subtraction to find out that they have 2 apples left.*

Math Facts

I can remember and use addition and subtraction facts up to 10.

- ***For example***, *I know that 4 + 3 = 7 and 7 − 3 = 4.*

Mental Math

I can use my brain to add and subtract numbers up to 20.

- ***For example***, *if I want to know what 9 + 5 is, I can count up from 9 and say 10, 11, 12, 13, 14.*

Addition and Subtraction

I can use pictures and objects to help add and take away numbers up to 50.

- ***For example***, *if I have 10 toy cars and my brother has 7 toy cars, we can put all the cars together and count them to find out that we have 17 cars in total.*

Multiplication and Division

I can solve problems where things are shared equally.

- ***For example***, *if I have 12 gummy bears and I want to share them equally with my 3 friends, I can count how many gummy bears each friend gets (4 gummy bears).*

Plus Zero Addition Strategy

Add 0 to a number and the number stays the same.

5 + 0 = 5

Add 0 to the number. Write the answer.

1 0 + 10 = _____

2 5 + 0 = _____

3 6 + 0 = _____

4 0 + 3 = _____

5 18 + 0 = _____

6 15 + 0 = _____

I can add 0 to a number.

Count On Addition Strategy

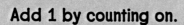

Add 1 by counting on.

4 + 1 = _____

Start with the greater number. Count on by 1.

4   5

Stop when 1 finger is up.

4 + 1 = **5**

Add 2 by counting on.

4 + 2 = _____

Start with the greater number. Count on by 2.

4   5   6

Stop when 2 fingers are up.

4 + 2 = **6**

Count on to add.

1   6 + 1 = ____

6, ____

2   4 + 2 = ____

4, ____, ____

3   12 + 1 = ____

12, ____

4   18 + 2 = ____

18, ____, ____

I can count on to add.

Using a Number Line to Add

Use a number line to add.

6 + 3 = __9__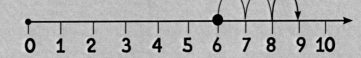

Mark a dot at 6. Draw 3 jumps to count on.
Stop at 9.

Use the number line to add. Mark a dot to show where to start. Next, count on by drawing the jumps. Write the answer.

1 3 + 4 = ____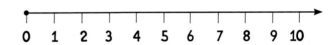

2 8 + 2 = ____

3 2 + 7 = ____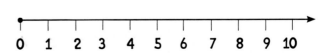

© Chalkboard Publishing Inc   I can use a number line to add.   49

Turn Around Addition Strategy

Add the numbers in any order and the total stays the same.

5 + 2 = 7

2 + 5 = 7

Use the ten frames to show adding numbers in two ways. Use two different colours. Then, write the answers.

1 7 + 2 = ____

2 + 7 = ____

2 5 + 4 = ____

4 + 5 = ____

3 3 + 6 = ____

6 + 3 = ____

I can add using the turn around strategy.

Counting Doubles Addition Strategy

Add the number to itself and that number doubles.

3 + 3 = 6

Add the number to itself to double it.

1
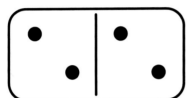

___ + ___ =

2
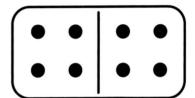

___ + ___ =

3
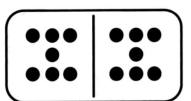

___ + ___ =

4
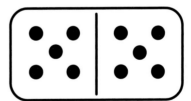

___ + ___ =

I can add the number to itself to double it.

Doubles Plus One Addition Strategy

Double the number and add one more to solve.

1.
3 + 3 = ____   so   3 + 4 = ____

2.
4 + 4 = ____   so   4 + 5 = ____

3.
5 + 5 = ____   so   5 + 6 = ____

4.
6 + 6 = ____   so   6 + 7 = ____

I can double the number and add 1.

Draw a Picture Addition Strategy

Draw circles to help you solve.

1

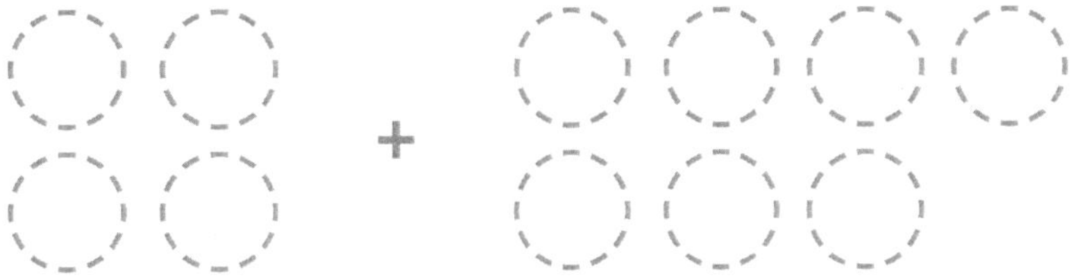

4 + 7 = ____

2

8 + 5 = ____

3

9 + 3 = ____

4

10 + 6 = ____

I can draw a picture to add.

Addition Word Problems

1 Draw a picture to help you solve the problem.

a Mia has 3 🚂s. Mia's mother gives her 1 more 🚂.
How many 🚂s does Mia have altogether?

_____ _____ = _____

b Kyle sees 6 🦁s at the zoo. 4 more 🦁s arrive.
How many 🦁s does Kyle see in total?

_____ _____ = _____

c Erica has 6 🌸s. She picks 9 more 🌸s.
How many 🌸s does Erica have in all?

_____ _____ = _____

I can add to solve a word problem.

Addition Word Problems (continued)

2 Draw a picture to help you solve the problem.

a There are **5** 🐦s in the bird bath. **9** more 🐦s arrive.
How many 🐦s are in the bird bath altogether?

____ ☐ ____ = ____

b Omar has **10** 📖s. His dad gives him **5** more 📖s.
How many 📖s does Omar have in total?

____ ☐ ____ = ____

c Tia brings **6** ⚽s to the park. **3** more ⚽s are added.
How many ⚽s does Tia have at the park?

____ ☐ ____ = ____

I can add to solve a word problem.

Addition Word Problems (continued)

3 Draw a picture to help you solve the problem.

a) There are **9** 🎺s in the box. **2** more 🎺s are added. How many 🎺s are in the box altogether?

_____ ☐ _____ = _____

b) Elijah washes **2** 🚗s. Elijah helps his dad wash **5** more 🚗s. How many 🚗s are washed in total?

_____ ☐ _____ = _____

c) Tania brings **7** 🌭s to the park. **2** more 🌭s are added. How many 🌭s does Tania have at the park?

_____ ☐ _____ = _____

I can add to solve a word problem.

Adding Ten More

1 Add the ten blocks.

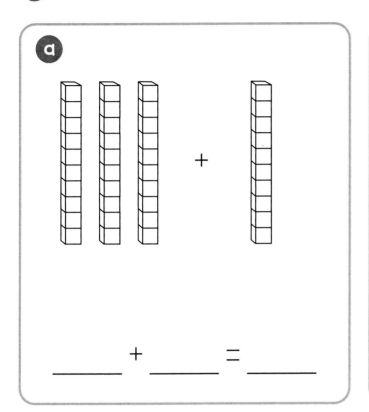

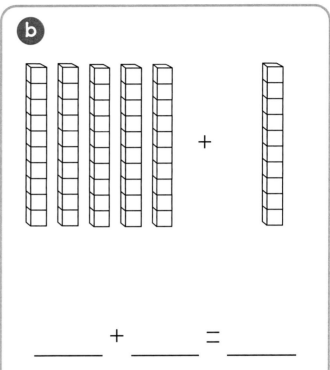

2 Add ten more to each number.

a 40 + 10 = ____

b 20 + 10 = ____

c 60 + 10 = ____

d 80 + 10 = ____

e 10 + 10 = ____

f 70 + 10 = ____

I can add ten blocks.

Addition: Sums to 10

Solve each addition sentence. Use the answer in the colour key. Use the key to colour the picture.

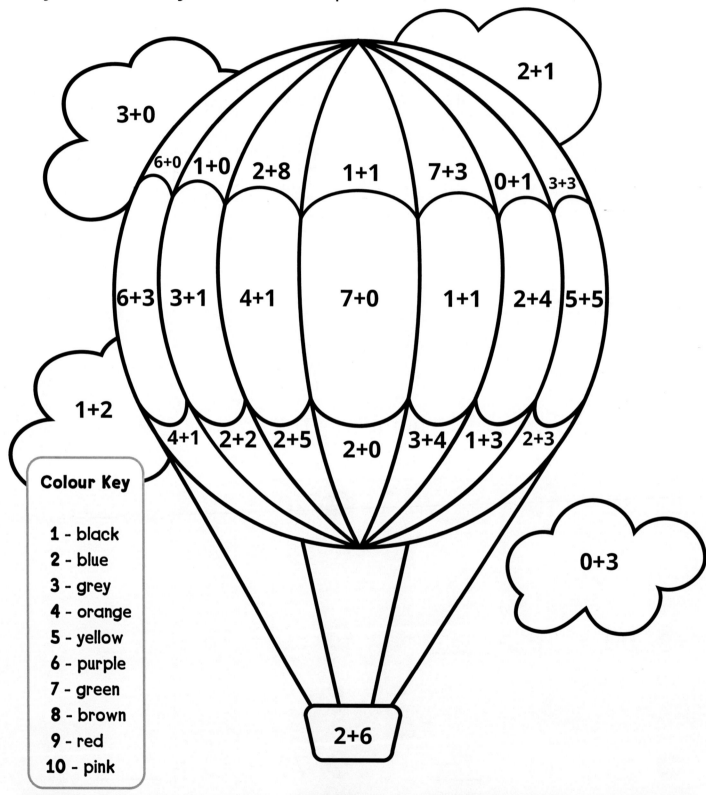

Colour Key

1 - black
2 - blue
3 - grey
4 - orange
5 - yellow
6 - purple
7 - green
8 - brown
9 - red
10 - pink

I can add sums to 10.

Addition: Sums to 20

1 Add. Use the number line or counters to help you add.

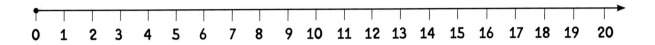

a) 5 + 8

b) 2 + 7

c) 6 + 6

d) 11 + 5

e) 7 + 4

f) 12 + 8

g) 10 + 2

h) 4 + 9

i) 15 + 2

j) 7 + 10

k) 1 + 9

l) 10 + 1

I can add sums to 20.

Minus Zero Subtraction Strategy

Subtract 0 from a number and the number stays the same.

8 – 0 = 8

Subtract zero from the number to solve.

❶ 5 – 0 = ____

❷ 6 – 0 = ____

❸ 7 – 0 = ____

❹ 13 – 0 = ____

❺ 9 – 0 = ____

❻ 20 – 0 = ____

I can subtract 0 from a number.

Counting Back Subtraction Strategy

Subtract 1 by counting back.

3 − 1 = _____

Count back from the first number. Count out loud.

3   2

Stop when 1 finger is down.

3 − 1 = **2**

Subtract 2 by counting back.

5 − 2 = _____

Count back from the first number. Count out loud.

5   4   3

Stop when 2 fingers are down.

5 − 2 = **3**

Subtract by counting back.

1 8 − 1 = _____

8, _____

2 6 − 2 = _____

6, _____, _____

3 7 − 1 = _____

7, _____

4 9 − 2 = _____

9, _____, _____

I can count back to subtract.

Using a Number Line to Subtract

Use a number line to subtract.

8 − 4 = __4__

Mark a dot at 8. Draw 4 jumps to count back. Stop at 4.

Use the number line to subtract.
Mark a dot to show where you start.
Next, count back by drawing the jumps.

1 7 − 3 = _____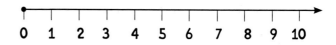

2 4 − 1 = _____

3 8 − 2 = _____

4 9 − 1 = _____

I can subtract using a number line.

A Number Minus Itself
Subtraction Strategy

Subtract a number from itself and the result is 0.

7 − 7 = 0

Subtract the number from itself to solve.

1) 8 − 8 = ____

2) 5 − 5 = ____

3) 10 − 10 = ____

4) 2 − 2 = ____

5) 6 − 6 = ____

6) 9 − 9 = ____

7) 15 − 15 = ____

8) 20 − 20 = ____

I can subtract a number from itself.

Doubles Subtraction Strategy

If you know the doubles fact, then you know the related subtraction fact.

$2 + 2 = 4$ or $4 - 2 = 2$

Use the doubles fact to subtract.

1

_6_ − _3_ = ___

2

___ − ___ = ___

3

___ − ___ = ___

4

___ − ___ = ___

64 — I can use doubles facts to subtract.

Draw a Picture Subtraction Strategy

Draw circles. Then cross them out to help you subtract.

1

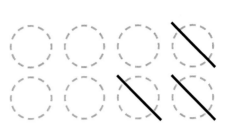

8 − 3 = _____

2

17 − 10 = _____

3

14 − 6 = _____

4

20 − 5 = _____

I can draw and cross out pictures to subtract.

Subtraction Word Problems

1 Solve the problem. Draw a picture to help you solve the problem.

a Amir has 10 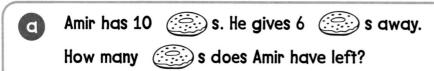s. He gives 6 🍩s away. How many 🍩s does Amir have left?

_____ ☐ _____ = _____

b Sonia has 20 s in all. Anita buys 10 🧁s from Sonia. How many 🧁s does Sonia have left?

_____ ☐ _____ = _____

c Roy has 5 . Marco has 4 🐟. How many more 🐟 does Roy have than Marco?

_____ ☐ _____ = _____

I can subtract to solve a word problem.

Subtraction Word Problems (continued)

2 Solve the problem. Draw a picture to help you solve the problem.

a) Dominic has 17 [B]s. Julian has 12 [B]s.

How many more [B]s does Dominic have than Julian?

_____ ◻ _____ = _____

b) Sarah makes 18 🥪s. She gives away 7 🥪s.

How many 🥪s does Sarah have left to give away?

_____ ◻ _____ = _____

c) 16 kids like 🍔s for dinner best.

7 kids like 🌭s for dinner best.

How many more kids like 🍔s for dinner than 🌭s?

_____ ◻ _____ = _____

I can subtract to solve a word problem.

Subtraction Word Problems (continued)

3 Solve the problem. Draw a picture to help you solve the problem.

a Jasper has 12 📖s. He donates 6 📖s.
How many 📖s does Jasper have left?

_____ ☐ _____ = _____

b Willow picks 10 ❀s. Willow gives Daisy 5 ❀s.
How many ❀s does Willow have left?

_____ ☐ _____ = _____

c Morgan has 5 ⚽s. Oscar has 4 ⚽s.
How many more ⚽s does Morgan have than Oscar?

_____ ☐ _____ = _____

I can subtract to solve a word problem.

Ten More, Ten Less

1 Add ten and subtract ten from the numbers.

10 Less	Number	10 More
	30	
	90	
	50	
	20	
	60	
	80	

I can subtract 10 from a number.

Subtracting Numbers from 0 to 10

Solve each subtraction sentence. Use the answer in the colour key. Use the key to colour the picture.

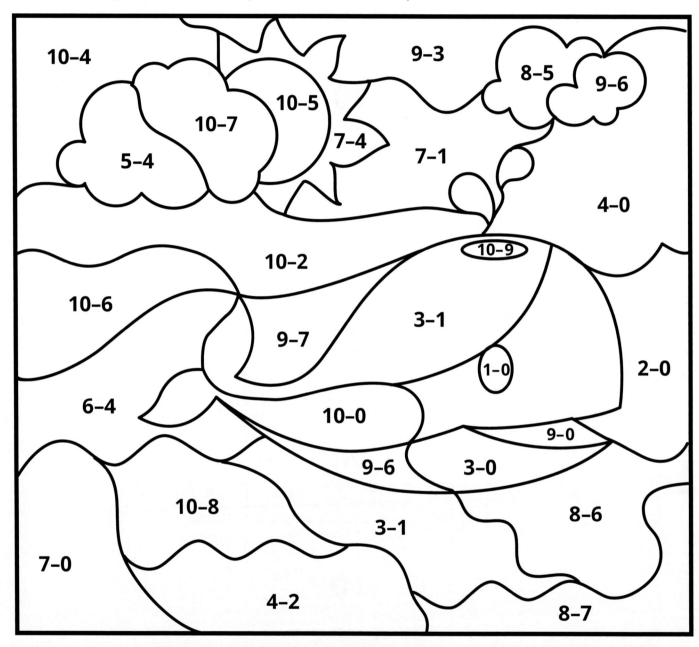

Colour Key

| 1 - black | 3 - grey | 5 - yellow | 7 - green | 9 - red |
| 2 - blue | 4 - orange | 6 - purple | 8 - brown | 10 - pink |

I can subract numbers from 0 to 10.

Subtracting Numbers from 0 to 20

Subtract. Use the number line to count back.

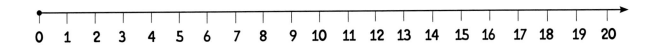

1) 7
   − 5
   ___

2) 10
   − 2
   ___

3) 14
   − 4
   ___

4) 11
   − 3
   ___

5) 19
   − 5
   ___

6) 12
   − 12
   ___

7) 11
   − 1
   ___

8) 15
   − 2
   ___

9) 13
   − 0
   ___

10) 17
   − 1
   ___

11) 10
   − 7
   ___

12) 16
   − 8
   ___

I can subtract numbers from 0 to 20.

Show What You Know — Sums from 0 to 10

2	0	2	3	4	2	3
+ 2	+ 2	+ 4	+ 2	+ 3	+ 1	+ 3

5	4	0	1	7	5	5
+ 2	+ 4	+ 1	+ 1	+ 2	+ 1	+ 3

1	2	3	9	10	0
+ 2	+ 3	+ 6	+ 1	+ 0	+ 0

Math Test Score ____ / 20

Show What You Know — Sums from 11 to 20

15	8	14	6	17	9	8
+ 5	+ 8	+ 6	+ 9	+ 2	+ 5	+ 9

13	9	16	6	9	10	11
+ 6	+ 7	+ 1	+ 6	+ 9	+ 5	+ 2

7	12	6	12	10	9
+ 6	+ 3	+ 5	+ 8	+ 4	+ 3

Math Test Score ____ / 20

Show What You Know — Subtraction from 0 to 10

9	4	10	5	8	7	9
− 4	− 2	− 4	− 2	− 0	− 3	− 7

6	8	6	3	0	10	5
− 4	− 6	− 3	− 3	− 0	− 6	− 3

7	1	9	2	5	10
− 5	− 1	− 5	− 0	− 4	− 5

Math Test Score
_____
20

Show What You Know — Subtraction from 11 to 20

13	11	19	16	13	11	20
− 6	− 4	− 9	− 9	− 3	− 7	− 10

16	15	12	19	14	17	18
− 0	− 5	− 8	− 7	− 3	− 6	− 9

11	14	13	18	17	20
− 5	− 6	− 1	− 5	− 2	− 6

Math Test Score
_____
20

Exploring Equal Groups

1 Fill in the blanks to find how many there are.

a) __5__ groups of _____ = _____

b) __2__ groups of _____ = __8__

c) _____ groups of __3__ = _____

I can explore equal groups.

Exploring Equal Groups (continued)

2 Fill in the blanks to find how many there are.

__2__ groups of _____ = __4__

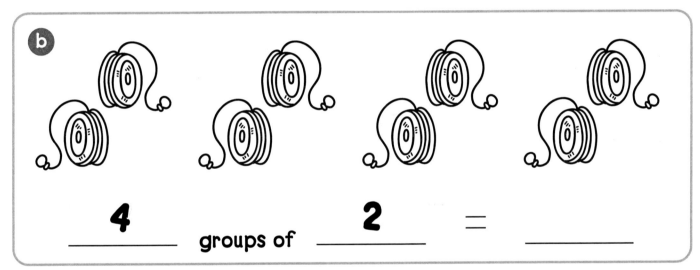

__4__ groups of __2__ = _____

__2__ groups of _____ = _____

Exploring Equal Groups

1 Fill in the blanks to find how many there are.

a) __3__ groups of _____ = __6__

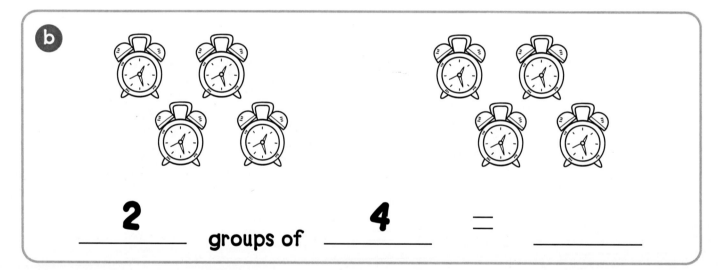

b) __2__ groups of __4__ = _____

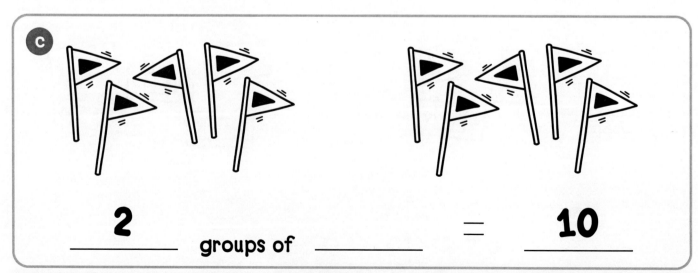

c) __2__ groups of _____ = __10__

I can explore equal groups.

Exploring Equal Groups (continued)

2 Fill in the blanks to find how many there are.

a

___3___ groups of _____ = _____

b

___2___ groups of _____ = ___6___

c

___2___ groups of _____ = ___10___

I can explore equal groups.

I Can Checklist:
Patterns and Relationships

Find and talk about patterns, even ones you see in everyday life. For example:

- *I can* look at different patterns around me, like patterns on a rug, patterns on clothes, or patterns in nature like stripes or spots.
- *I can* talk about what I see in the patterns, like noticing that they repeat or that they have a specific order.

Make patterns using moves, sounds, things, shapes, letters, and numbers. For example:

- *I can* make patterns with my body, in a specific order, like clap, jump, clap.
- *I can* make patterns with sounds, like clapping hands, tapping on a table.
- *I can* make patterns with objects, like car, doll, car, doll.
- *I can* make patterns with shapes, like drawing a triangle, circle, triangle.
- *I can* make patterns with letters and numbers, like ABCABC or 123123.

Figure out the rules of patterns. Use them to continue the pattern, guess what comes next, and find what's missing. For example:

- *I can* figure out the rule of a pattern.
- *I can* continue a pattern, like knowing that a pattern that starts with 1, 3, 5 will continue with 7, 9, and so on.
- *I can* predict what comes next in a pattern, based on the rule.
- *I can* find missing parts in a pattern, like knowing that a pattern goes red, blue, ___, red, blue, green, and figuring out that the missing part is green.

Make and talk about patterns to show how numbers up to 50 are connected. For example:

- *I can* make patterns with numbers, like counting by 2s, 5s, or 10s in a specific order, like 2, 4, 6, 8, or 10, 20, 30, 40.
- *I can* talk about the relationship between the numbers in the pattern, like noticing that they are getting bigger or that they have a special pattern.

Patterns Are All Around Us!

A pattern repeats over and over.

1 Be a detective and look for patterns around you every day! In each box, glue pictures, draw your own pictures, or write about the patterns you find around where you live.

a) Patterns I See Inside
   Some ideas: • wallpaper • furniture • tiles • daily routines

b) Patterns I See in Nature
   Some ideas: • leaves • butterflies • animal fur • seasonal cycles

I can find patterns all around me.

Patterns Are All Around Us! (continued)

2 Circle and colour the patterns you see.

a Colour the patterns on **animals** [red].

b Colour the patterns on **fruits and vegetables** [green].

c Colour the patterns on **human-made items** [blue].

I can identify patterns on objects.

Patterns in Everyday Life

A **pattern** repeats over and over.

1 Circle the core of each pattern that happens in nature.

a **The Seasons**

b **Day and Night**

2 What patterns do you notice in your everyday life?

I can identify patterns in everyday life.

Extend the Pattern

A **pattern** repeats.

The **core** is the smallest part of a pattern that repeats.

1 Circle the core of the repeating pattern. Extend the pattern.

a) A B A B A B ___ ___ ___

b) ↑ ↑ ↓ ↑ ↑ ↓ ___ ___ ___

c) ♥ ♥ ♥ ♥ ♥ ♥ ___ ___ ___

d) ▢ ▨ ◯ ◯ ▨ ◯ ___ ___ ___

I can extend patterns.

Extend the Pattern (continued)

2 Circle the core of the repeating pattern. Extend the pattern.

a)

b) _____

c)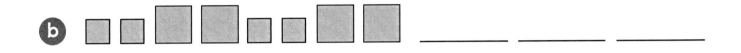

d) → → ↓ → _____

e) A B C A B C _____

I can extend patterns.

Growing Patterns

A **growing pattern** gets bigger by the same amount each time.

1 Extend the **growing** number pattern. Describe what you counted by.

a 1 2 3 4 5 6 ____ ____ ____

Start at _____ then add _____ each time.

b 2 4 6 8 10 ____ ____ ____

Start at _____ then add _____ each time.

2 Extend the **growing** shape pattern.

_____ _____

I can extend the growing number pattern.

Shrinking Patterns

A **shrinking pattern** gets smaller by the same amount each time.

1 Extend the **shrinking** number pattern.

a **10 9 8 7 6** ____ ____ ____

b **20 19 18 17** ____ ____ ____

b **14 13 12 11** ____ ____ ____

2 Extend the **shrinking** shape pattern.

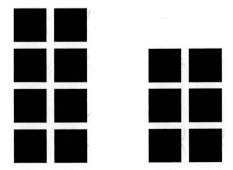 _____ _____

How Does the Pattern Attribute Change?

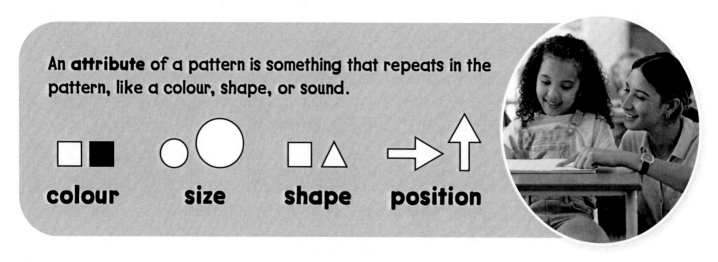

An **attribute** of a pattern is something that repeats in the pattern, like a colour, shape, or sound.

colour   size   shape   position

1 How does the pattern **attribute** change?
Circle **size**, **colour**, **shape**, or **position**.

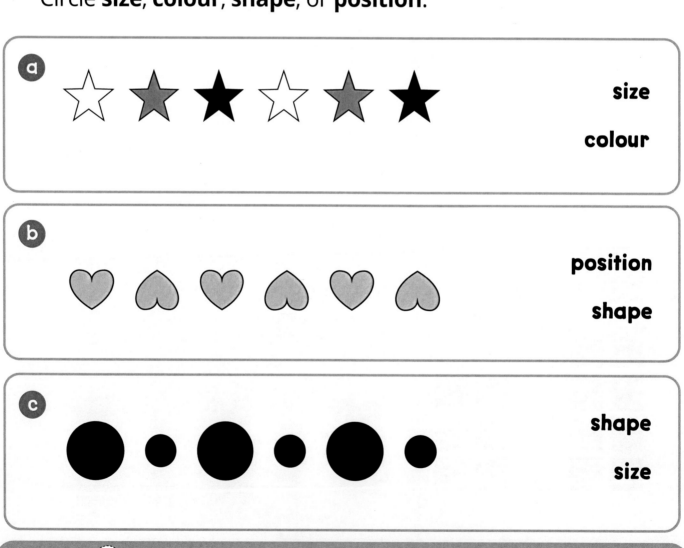

a) size / colour

b) position / shape

c) shape / size

I can identify the pattern attribute that changes.

How Does the Pattern Attribute Change? (continued)

2 How does the pattern **attribute** change?
Circle **size**, **colour**, **shape**, or **position**.

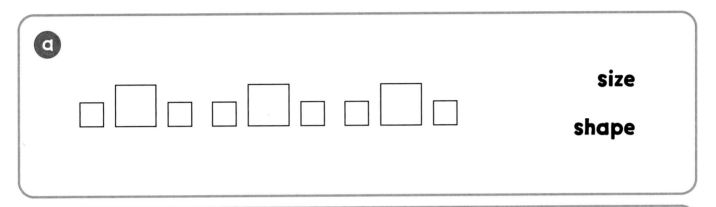

a) size / shape

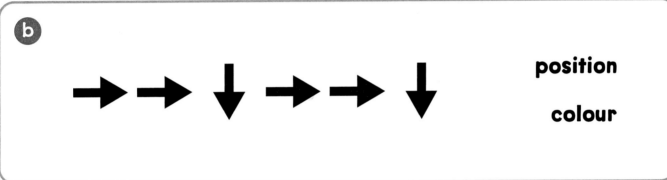

b) position / colour

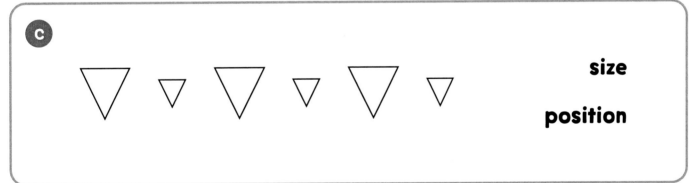

c) size / position

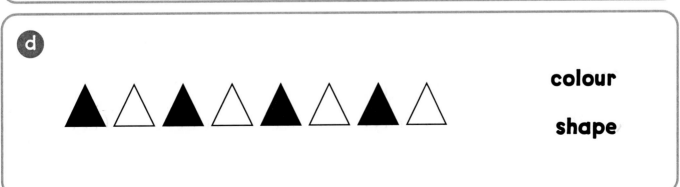

d) colour / shape

I can identify the pattern attribute that changes.

Describing the Pattern Rule

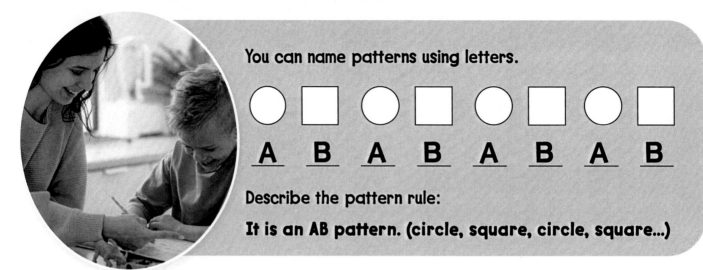

You can name patterns using letters.

○ □ ○ □ ○ □ ○ □
A B A B A B A B

Describe the pattern rule:

It is an AB pattern. (circle, square, circle, square...)

Name the pattern using letters.
Describe the pattern rule to a partner.

1
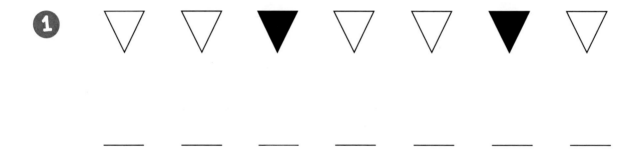

2
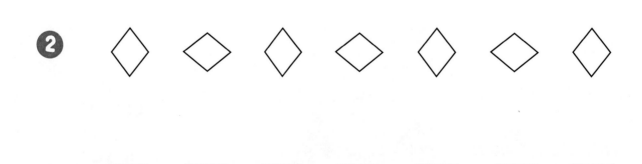

I can describe the pattern rule.

Identifying and Describing Patterns

Circle the pattern core. Then write the pattern rule for the repeating pattern.

| AAB | ABC | AABB | ABBC |

1.      _____

2.      _____

3.  _____

4.   _____

I can identify and describe the pattern rule.

Complete the Pattern

1 Complete the missing part of the pattern.

a)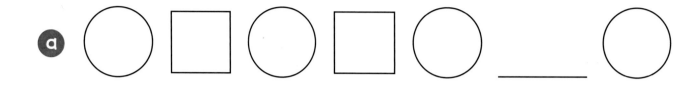

b) **50   45   40 ____ ____ 25**

c)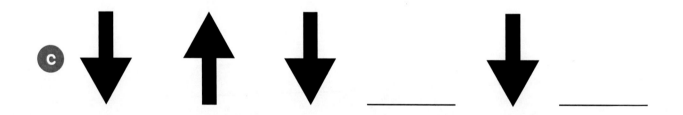

d) ○ ◯ ○ ◯ ○ ◯ ____ ____

Complete the Pattern (continued)

2 Complete the missing part of the pattern.

a)

b)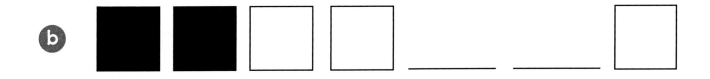

c)

d) 2   4   6 ____ 10 12 ____

I can complete the pattern.

Create a Pattern

1 Colour an **AAB** pattern. Circle the core of the pattern.

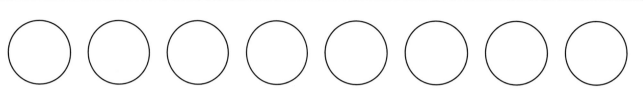

2 Create a pattern that shows a **shape change**.

3 Create a pattern that shows a **size change**.

4 Create a pattern that shows a **position change**.

I can create a pattern.

Create a Pattern Rule

1 Colour in a pattern. Circle the core of the pattern.

◯ ◯ ◯ ◯ ◯ ◯ ◯ ◯

Describe your pattern rule.
_____
_____
_____

2 Create a pattern. Circle the **core** of the pattern.

Describe your pattern rule.
_____
_____
_____

I can create a pattern rule.

Skip Counting to 100

1 Count by **2s**. Use **yellow** to colour each number you count.
Count by **5s**. Draw a **circle** around each number you count.
Count by **10s**. Use **blue** to colour each number you count.

1	2	3	4	5	6	7	8	9	10
11	12	13	14	15	16	17	18	19	20
21	22	23	24	25	26	27	28	29	30
31	32	33	34	35	36	37	38	39	40
41	42	43	44	45	46	47	48	49	50
51	52	53	54	55	56	57	58	59	60
61	62	63	64	65	66	67	68	69	70
71	72	73	74	75	76	77	78	79	80
81	82	83	84	85	86	87	88	89	90
91	92	93	94	95	96	97	98	99	100

I can skip count to 100.

Line Patterns

Fill the sections in with different line patterns. Then colour in your picture.

More and Fewer Objects

1 Draw to show **more**, **equal**, or **fewer** shapes. Colour the shapes.

a Draw a set with **1 more** circle.

b Draw a set with **1 fewer** triangle.

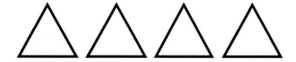

c Draw a set with an **equal number** of triangles.

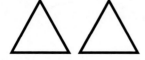

I can show more, equal, or fewer shapes.

More and Fewer Objects (continued)

2 Draw to show **more**, **equal**, or **fewer** shapes. Colour the shapes.

a) Draw a set with an **equal number** of triangles.

b) Draw a set with **2 more** squares.

c) Draw a set with **2 fewer** circles.

I can show more, equal, or fewer shapes.

Show What You Know!

1 Circle the core of the repeating pattern. Extend the pattern.

2 Complete the missing part of the pattern.

3 Extend the **shrinking** number pattern.

14   13   12   11 ____ ____ ____

4 Extend the **growing** shape pattern.

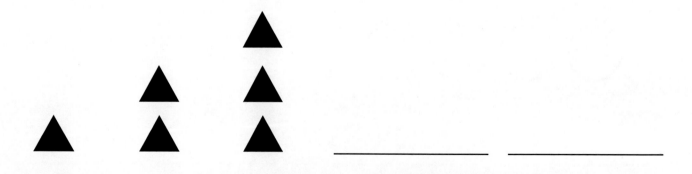

Show What You Know! (continued)

5 Write the pattern rule for the repeating pattern.

AAB   ABC   ABBC

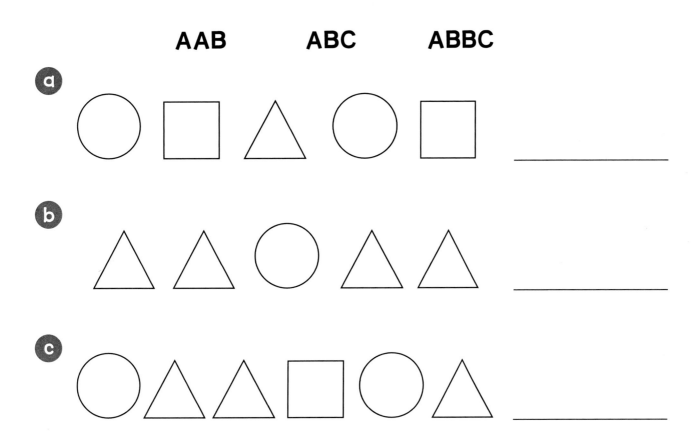

6 How does the pattern **attribute** change?
Circle **size**, **colour**, **shape**, or **position**.

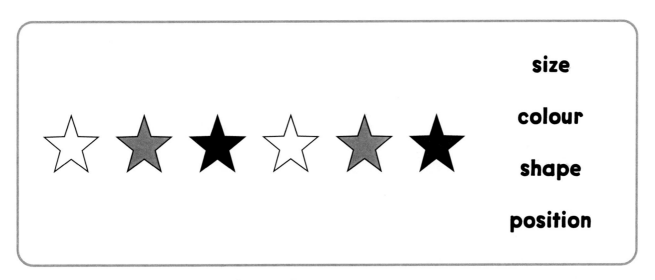

size

colour

shape

position

I Can Checklist:
Variables, Equalities, and Inequalities

Variables

Find things in real life that can change in amount and things that always stay the same. For example:

- *I can* look at things around me and see which things can change, like the number of apples in a basket or the temperature outside.

- *I can* also see which things always stay the same, like the color of my hair or the number of fingers I have.

Equalities and Inequalities

Figure out if pairs of adding and subtracting problems give the same answer or not. For example:

- *I can* look at addition and subtraction expressions, like 2 + 3 and 4 + 1, and see if they are the same or different.

- *I can* figure out if they are equal (the same) or not equal (different).

Find and use numbers up to 50 that mean the same thing in different situations. For example:

- *I can* find equivalent relationships between different numbers up to 50.

- *I can* know that 5 + 3 is the same as 4 + 4 or 2 + 6.

- *I can* use these equivalent relationships to help me solve problems or make calculations.

Understanding Things That Stay the Same and Things That Can Change

Sometimes things always stay the same. We call these things **constants**.

For example: There are always 24 hours in a day.

But there are other things that can change. We call these things **variables** because they can be different.

For example: The time you eat a snack.

Colour the things that stay the same ⬚ orange ⬚.

Colour the things that can change ⬚ green ⬚.

| number of hours spent playing | 24 hours in a day |

| 12 months in a year | number of cents to buy something |

| number of months until a special event | number of cents in a dollar |

© Chalkboard Publishing Inc

Balance It!

1. Draw blocks to make the balance equal on both sides.

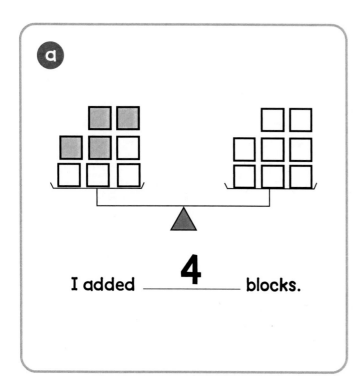

I added __4__ blocks.

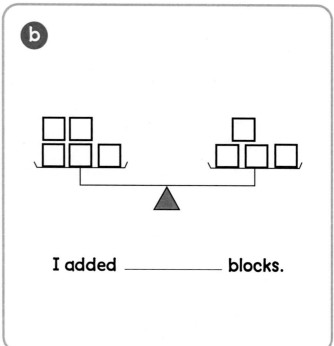

I added _____ blocks.

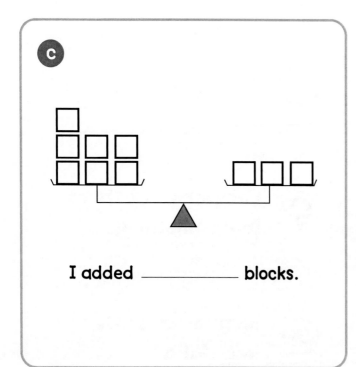

I added _____ blocks.

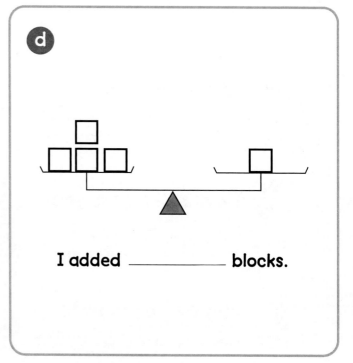

I added _____ blocks.

Balance It! (continued)

2. Use an **X** to take away blocks to make the balance equal on both sides.

a

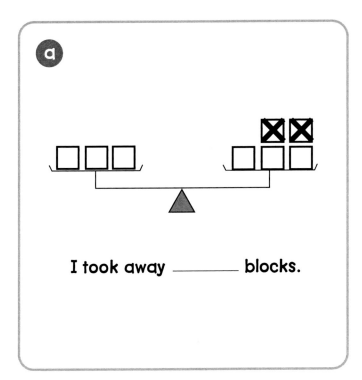

I took away _____ blocks.

b

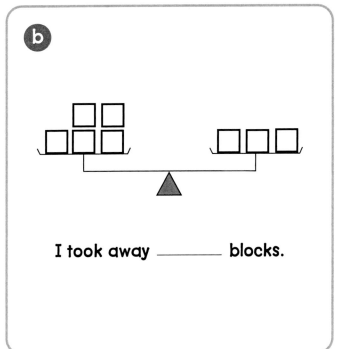

I took away _____ blocks.

c

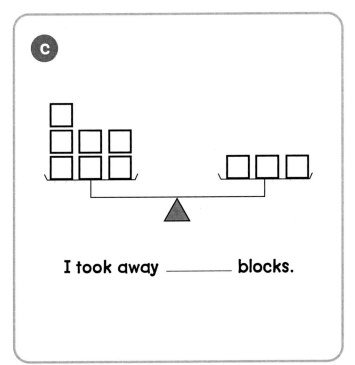

I took away _____ blocks.

d

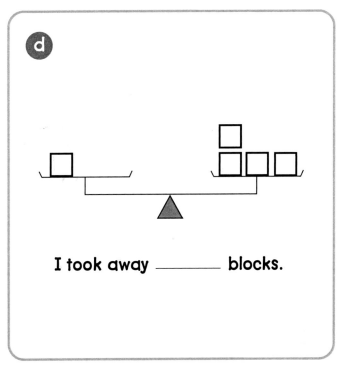

I took away _____ blocks.

I can take away blocks to balance the scale.

Making Addition Sentences

1 Show three ways to make each number. Use two colours.

a

___ + ___ = 6

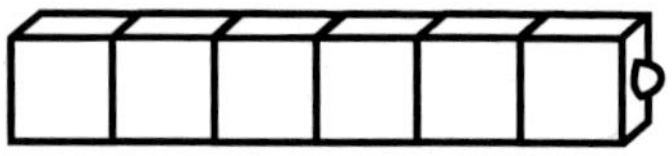

___ + ___ = 6

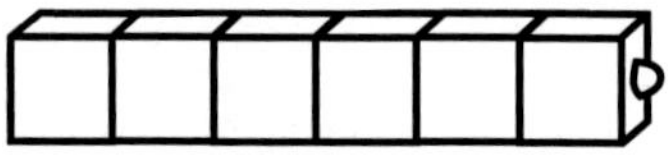

___ + ___ = 6

b

___ + ___ = 10

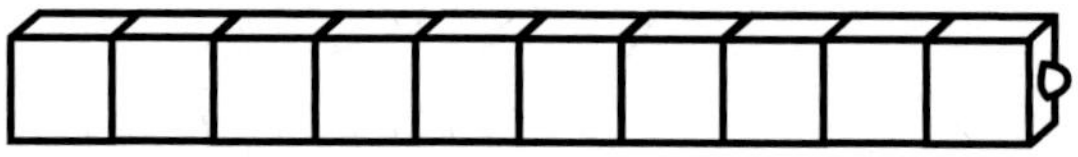

___ + ___ = 10

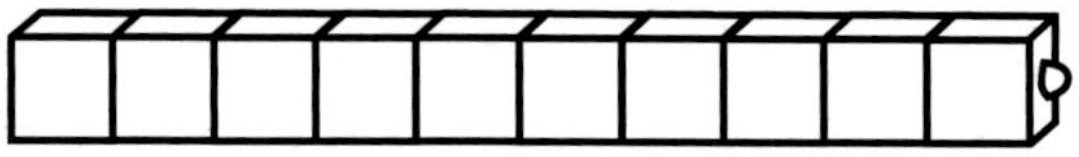

___ + ___ = 10

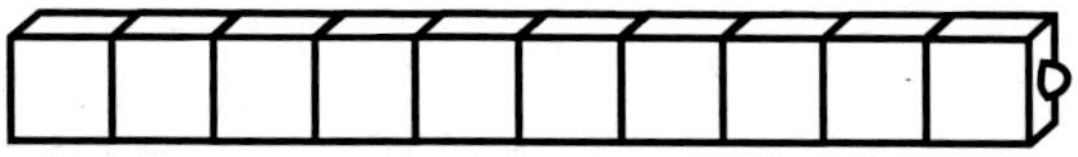

c

___ + ___ = 4

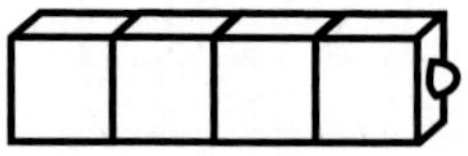

___ + ___ = 4

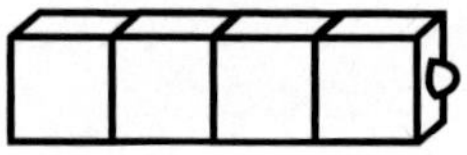

___ + ___ = 4

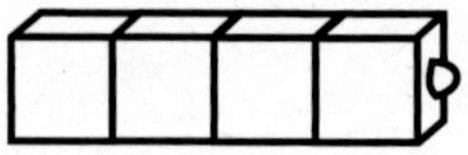

I can make addition sentences.

Making Addition Sentences (continued)

2 Show three ways to make each number. Use two colours.

a

____ + ____ = 8

____ + ____ = 8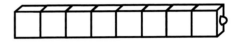

____ + ____ = 8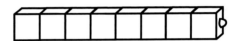

b

____ + ____ = 9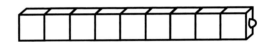

____ + ____ = 9

____ + ____ = 9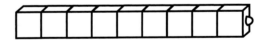

c

____ + ____ = 5

____ + ____ = 5

____ + ____ = 5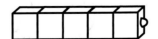

I can make addition sentences.

Making Subtraction Sentences

1 Cross out the blocks you want to take away. Colour the blocks that are left. Complete the subtraction sentence.

a

4 − ____ = ____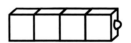

4 − ____ = ____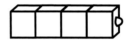

4 − ____ = ____

b

8 − ____ = ____

8 − ____ = ____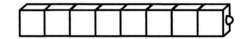

8 − ____ = ____

c

10 − ____ = ____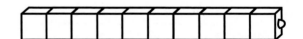

10 − ____ = ____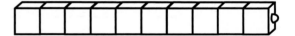

10 − ____ = ____

I can make subtraction sentences.

Making Subtraction Sentences (continued)

2 Cross out the blocks you want to take away. Colour the blocks that are left. Complete the subtraction sentence.

a

7 - ____ = ____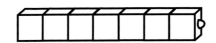

7 - ____ = ____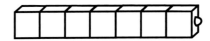

7 - ____ = ____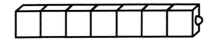

b

5 - ____ = ____

5 - ____ = ____

5 - ____ = ____

c

3 - ____ = ____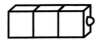

3 - ____ = ____

3 - ____ = ____

I can make subtraction sentences.

Number Fact Families

1. Read the numbers in the group.
   Add or subtract using the three numbers.

a ③ ⑤ ⑧

___ + ___ = ___

___ + ___ = ___

___ − ___ = ___

___ − ___ = ___

b ④ ③ ⑦

___ + ___ = ___

___ + ___ = ___

___ − ___ = ___

___ − ___ = ___

I can add and subtract to make number fact families.

Number Fact Families (continued)

2 Read the numbers in the group.
Add or subtract using the three numbers.

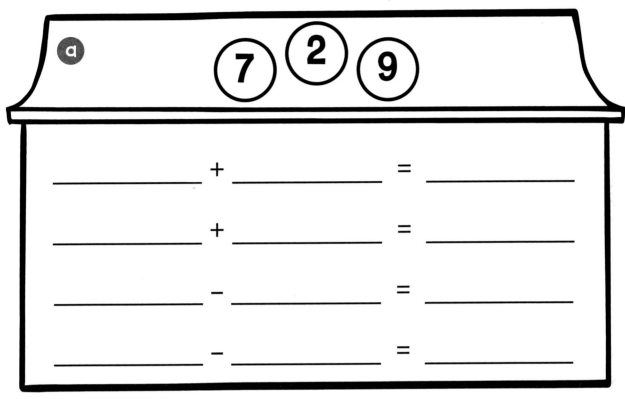

a 7 2 9

_____ + _____ = _____

_____ + _____ = _____

_____ − _____ = _____

_____ − _____ = _____

b 1 3 4

_____ + _____ = _____

_____ + _____ = _____

_____ − _____ = _____

_____ − _____ = _____

I can add and subtract to make number fact families.

Is It Equal?

1 Add the numbers on each side of the equal sign. If the answers are the same for both sides, draw a check mark (✓). If the answers are different, draw a slash (≠) through the equal sign. The first one is done for you.

a

$4 + 3 = 2 + 5$

$4 + 3 = \underline{7}$

$2 + 5 = \underline{7}$

b

$5 + 1 = 2 + 2$

$5 + 1 = \underline{}$

$2 + 2 = \underline{}$

c

$3 + 6 = 5 + 1$

$3 + 6 = \underline{}$

$5 + 1 = \underline{}$

d

$8 + 2 = 7 + 3$

$8 + 2 = \underline{}$

$7 + 3 = \underline{}$

I can solve and check addtion word problems.

Is It Equal? (continued)

2 Subtract the numbers on each side of the equal sign. If the answers are the same for both sides, draw a check mark (✓). If the answers are different, draw a slash (≠) through the equal sign.

a

8 − 2 = 9 − 3

8 − 2 = ___

9 − 3 = ___

b

10 − 5 = 4 − 3

10 − 5 = ___

4 − 3 = ___

c

6 − 4 = 3 − 0

6 − 4 = ___

3 − 0 = ___

d

10 − 6 = 7 − 3

10 − 6 = ___

7 − 3 = ___

I can solve and check addtion word problems.

Show What You Know!

1 Draw blocks or take away blocks to make the balance equal on both sides.

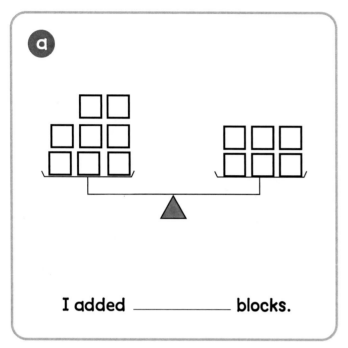

a

I added _____ blocks.

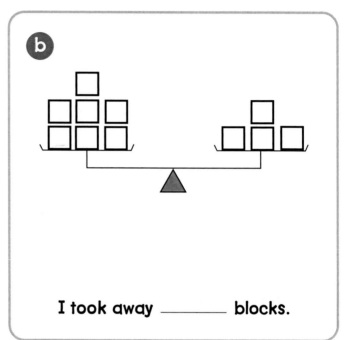

b

I took away _____ blocks.

2 Solve each side of the equal sign. If the answers are the same for both sides, draw a check mark (✓). If the answers are different, draw a slash (≠) through the equal sign.

a

5 + 4 = 7 + 1

5 + 4 = ___

7 + 1 = ___

b

8 − 4 = 6 − 2

8 − 4 = ___

6 − 2 = ___

Show What You Know! (continued)

3 Show three ways to make each number. Use two colours.

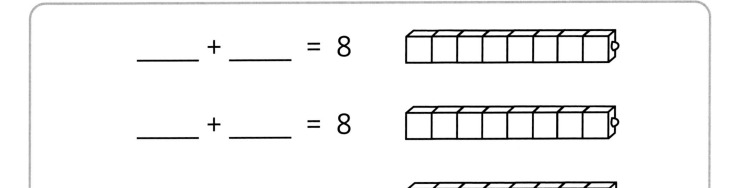

4 Read the numbers in each group.
Add or subtract using the three numbers.

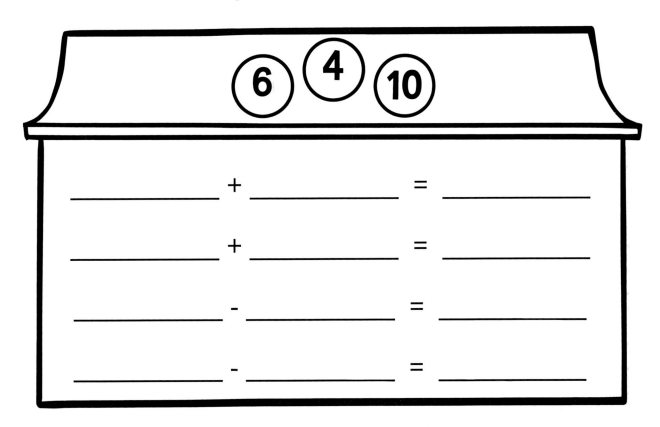

I Can Checklist:
Coding Skills

Exploring writing and executing code, including code that involves sequential events. For example:

- **I can** use coding to solve math problems or create representations of math situations.
- **I can** write code to count how many apples are in a basket or to show a number line on the computer screen.
- **I can** make things happen in a specific order, like having the computer show the numbers from 1 to 10 one by one.

Read and alter existing code, including code that involves sequential events, and describe how changes to the code affect the outcomes. For example:

- **I can** look at code that someone else wrote and understand what it does.
- **I can** make changes to the code, like changing the numbers or the order of events, and see how it affects the outcome.
- **I can** explain what will happen differently when I make changes to the code.

Coding in Everyday Life

Coding is like giving the computer a set of instructions in a language it understands, one step at a time.

For example, these things need coding to run:

washing machine   television   cell phone

Colour the pictures of things that need coding to run.

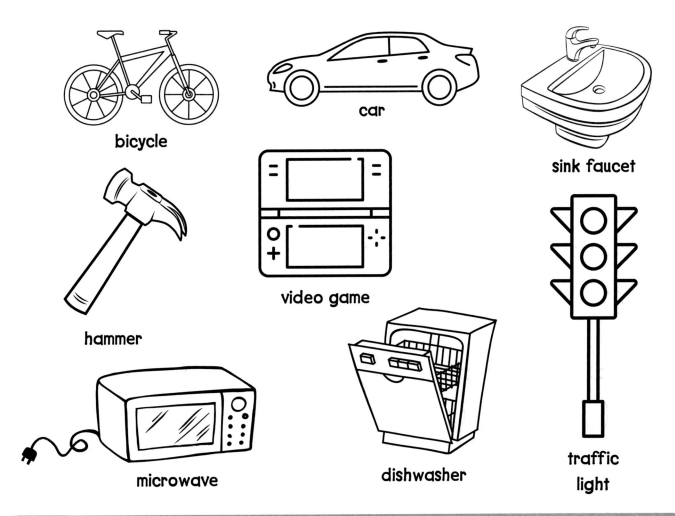

bicycle   car   sink faucet   hammer   video game   microwave   dishwasher   traffic light

I can identify things that need coding to run.

Read the Code

Code: the language we use to give a computer step-by-step instructions or commands.

	START HERE				
			🍴		
🛠️				🪣	

1 Begin on the "Start Here" square.
Follow the code and circle the object you land on.

a) → ↓ ↓ ← ←   🪣   🛠️   🍴

b) → → → ↓ ↓   🪣   🛠️   🍴

I can read code.

Read the Code (continued)

2. Begin on the "Start Here" square.
Follow the code and circle the object you land on.

a)

b)

I can read code.

Exploring Writing Code

Use arrow to show lines of code to move the car to the store.

For example:

Move right ➡
Move left ⬅
Move down ⬇
Move up ⬆

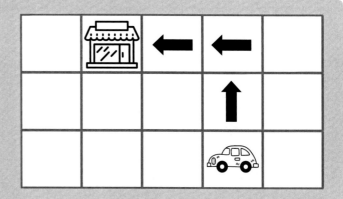

① Draw arrows in the squares to move the truck to the store.

a)

b)

I can explore and write code.

Exploring Writing Code (continued)

2 Draw arrows in the squares to move the rooster to the chick.

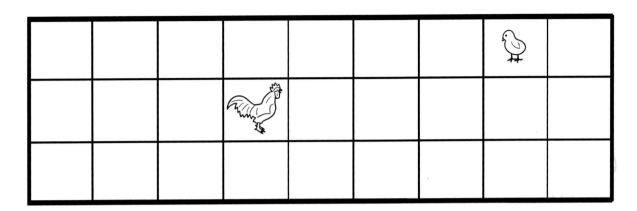

Draw arrows on the line to show the path.

_____

3 Draw arrows in the squares to move the player to the ball.

Draw arrows on the line to show the path.

_____

I can explore and write code.

Alter the Code

The word *alter* means to change. *Alter the code* means to write the same code in a different way or in a different order.

1 Begin on the "Start Here" square. Use the code to reach the 🐋. Draw the arrows on the grid.

code: ↑ ← ← ←

2 Alter the code on the line so that you get to the 🐋 in a different way. Write the new code below and draw the arrows on the grid.

_____

I can examine and alter code.

Alter the Code (continued)

3) Begin on the "Start Here" spot. Use the code to reach the 🪣. Draw the arrows on the grid.

code: ➡ ⬇ ⬇ ➡ ➡

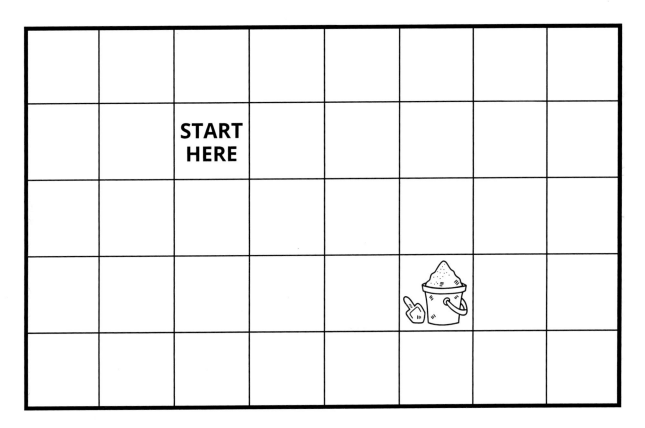

4) Alter the code on the line so that you get to the 🪣 in a different way. Write the new code below and draw the arrows on the grid.

_____

I can examine and alter code.

Show What You Know!

1 Circle and colour the objects that need coding to run.

wrench   computer   digital clock

2 Draw arrows in the squares to move the car to the store.

a

b

Show What You Know! (continued)

3 Begin on the "Start Here" spot. Use the code to reach the 🚩. Draw the arrows on the grid.

code: ➡ ⬇ ⬇ ➡ ➡

4 Alter the code on the line so that you get to the 🚩 in a different way. Write the new code below and draw the arrows on the grid.

_____

I Can Checklist:
Money Concepts

Learn about Canadian coins up to 50 cents and coins and bills up to $50, and see how their values are different. For example:

- *I can* recognize different coins in Canada, like the nickel, dime, quarter, loonie, and toonie.

- *I can* understand the value of each coin, like knowing that a nickel is worth 5 cents, a dime is worth 10 cents, and so on.

- *I can* also recognize different bills, like the $5 bill, $10 bill, and $20 bill, and understand their value.

Parent Tips:

Let's Play Shop:
How about turning learning into a game? Set up a pretend shop with your child's favorite toys, each with its own price tag. Give them a mix of coins and bills and let your child 'buy' stuff. It's a great way to help your child understand how to use different coins and bills.

Sorting Time:
Pour a handful of coins on the table and let's sort them together! This can be a fun and easy way to help your child recognize nickels, dimes, quarters, loonies, and toonies.

Money Show and Tell:
Grab a coin or bill and show it to your child. Talk about its value and why, even though a nickel is bigger than a dime, a dime is worth more.

Flashcards Fun:
You can make flashcards together! Write an amount of money on one side (like $1.25), and on the other side, show the coins that make up that amount.

Making Change:
When your child is ready, let's talk about making change. Start with something simple and increase the difficulty as your child gets the hang of it.

Canadian Coins

This is a nickel.

A nickel is worth 5¢ or 5 cents.

It has a beaver on one side.

Colour the nickel [silver].

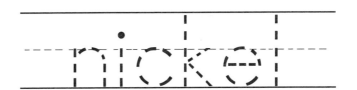

This is a dime.

A dime is worth 10¢ or 10 cents.

It has the Bluenose schooner on one side.

Colour the dime [silver].

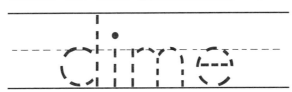

This is a quarter.

A quarter is worth 25¢ or 25 cents.

It has a caribou on one side.

Colour the quarter [silver].

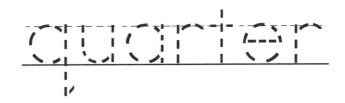

I can identify coins.

Canadian Coins (continued)

This is a loonie.

A loonie is worth 100¢, or 100 cents, or $1.00.

A loonie has a loon on one side.

Colour the loonie gold.

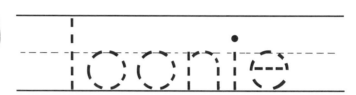

This is a toonie.

A toonie is worth 200¢, or 200 cents, or $2.00.

A toonie has a polar bear on one side.

Colour the outside silver.

Colour the inside gold.

I can identify coins.

Matching Canadian Coins

Match the coin to the correct amount.

     •   • 200¢

     •   • 25¢

   •   • 5¢

   •   • 10¢

   •   • 100¢

I can match coins.

Counting Nickels

A **nickel** is worth 5¢.
Count the nickels by 5s to find the value.

_____¢ _____¢ _____¢ = _____¢

_____¢ _____¢ _____¢ _____¢ _____¢ _____¢ = _____¢

_____¢ _____¢ _____¢ _____¢ _____¢ = _____¢

_____¢ _____¢ _____¢ _____¢ _____¢ _____¢ _____¢ = _____¢

I can count Nickels.

Counting Dimes

A **dime** is worth 10¢.
Count the dimes by 10s to find the value.

1)

____¢ ____¢ ____¢ = _____¢

2)

____¢ ____¢ ____¢ ____¢ = _____¢

3)

____¢ ____¢ ____¢ ____¢ ____¢ ____¢ = _____¢

4)

____¢ ____¢ ____¢ ____¢ ____¢ ____¢ ____¢ = _____¢

I can count dimes.

Comparing Money

Count the money in each set. Write the amount.
Print the correct symbol in the box between the sets.

| greater than | less than | equal to |

1

2

3

I can count and compare money.

Comparing Coin Sizes

Compare the coins in the set. Follow the instructions in the set.

1 Circle the **largest** coin.

2 Circle the **smallest** coin.

3 Order the coins from **largest** to **smallest**. Write the order below each.

_____   _____   _____   _____

I can compare coin sizes.

Canadian Bills

**This is a
5 dollar bill.**

This bill is worth $5 or 5 dollars.

It has Sir Wilfrid Laurier on one side.

Colour the bill 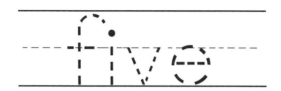.

five

**This is a
10 dollar bill.**

This bill is worth $10 or 10 dollars.

It has Viola Desmond on one side.

Colour the bill .

ten

**This is a
20 dollar bill.**

This bill is worth $20 or 20 dollars.

It has Queen Elizabeth II on one side.

Colour the bill 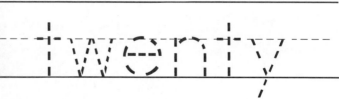.

twenty

I can identify bills.

Canadian Bills (continued)

This is a 50 dollar bill.

This bill is worth $50 or 50 dollars.

It has William Lyon Mackenzie King on one side.

Colour the bill red.

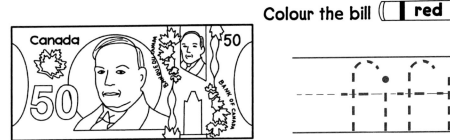

This is a 100 dollar bill.

This bill is worth $100 or 100 dollars.

It has Sir Robert Borden on one side.

Colour the bill brown.

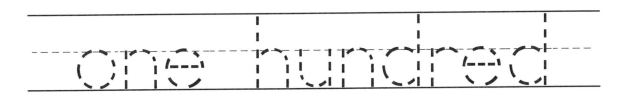

I can identify bills.

Comparing Bill Values

Write the value of each bill.
Print the correct symbol in the box.

| greater than | less than | equal to |

1

$_____ $_____

2

$_____ $_____

3

$_____    $_____

I can identify and compare bills.

Order the Coins and Bills

Order the coins and bills from **greatest** to **least** value.

Show What You Know!

1 A **nickel** is worth **5¢**.
Count the nickels by 5s to find the value.

       = _____ ¢

2 A **dime** is worth **10¢**.
Count the dimes by 10s to find the value.

      = _____ ¢

3 A **quarter** is worth **25¢**.
Count the quarters by 25s to find the value.

    = _____ ¢

4 A **loonie** is worth **$1**.
Count the loonies by 1s to find the value.

     = $ _____

Show What You Know! (continued)

5 Count the money in each set. Write the amount.
Print the correct symbol in the box between the sets.

| greater than | less than | equal to |

a

_____   _____

b

_____   _____

I Can Checklist:
Data Literacy

Data Collection and Organization

✓ Sort things into groups based on one characteristic. For example:
- **I can** group objects or people based on a specific attribute, like sorting toys by colour.
- **I can** explain the rules I used to sort, like saying I put all the red toys together.

✓ Collect and put information in tally tables to keep it organized. For example:
- **I can** learn new things by watching and asking questions.
- **I can** write down or draw what I learn.
- **I can** put the information in tally tables to keep it organized.

Data Visualization

✓ Show information in an organized way with titles and labels. For example:
- **I can** make graphs and pictures to show information using pictures or symbols.
- **I can** make sure to use a title and other labels to tell what the graph is about.

Data Analysis

✓ Put things in order from most to least frequent. For example:
- **I can** look at the information in charts, graphs, or pictures, and see which group shows up the most or has the biggest number.
- **I can** put the groups in order from the one that shows up the most to the one that shows up the least.

✓ Look closely at different groups of information shown in different ways. For example:
- **I can** look at the information and ask questions, like "Which group has the most?"
- **I can** use the information to answer my questions and understand what it means.
- **I can** explain my ideas and share what I think, based on the information I see.

Create Your Own Survey!

What is your favourite _____?

- pet
- colour
- fruit
- home activity
- snack
- season
- winter activity
- summer activity
- sport
- recess activity
- pizza topping
- ice cream flavour
- dessert
- restaurant
- meal
- day of the week
- superhero
- author
- reading genre
- music genre
- school subject
- holiday
- cereal
- breakfast meal
- game
- coin
- cartoon
- lunch meal
- vegetable
- time of day

Make your own questions...

- What do you prefer?
- What do you like best?
- What is your estimation?
- What is your prediction?

More ideas:

- How many people are in your family?
- What colour hair do you have?
- What colour eyes do you have?
- Would you rather live in the rainforest or the ocean?

Exploring Pictographs

A **pictograph** is a way of using pictures to show data.

Data is information or facts that we can collect to learn new things. It can be numbers, pictures, or words

1 Read the pictograph to answer the questions.

What is your favourite party food?

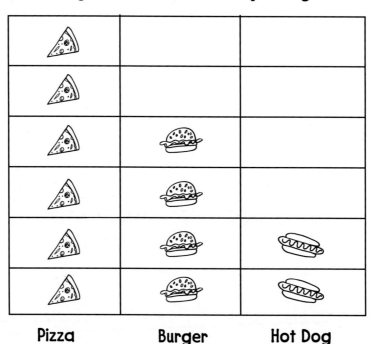

Pizza   Burger   Hot Dog

a How many votes? _____ _____ _____

b Circle the **most popular** food.

c Circle the **least popular** food.

I can read and understand pictographs.

Exploring Pictographs (continued)

2 Read the tally chart to answer the questions.

What is your favourite playground activity?

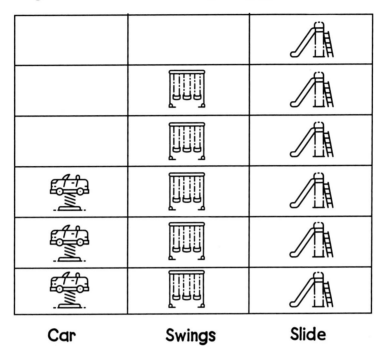

a How many votes? 🚗 ____ 🎏 ____ 🛝 ____

b Circle the **most popular** activity.   🚗   🎏   🛝

c Circle the **least popular** activity.   🚗   🎏   🛝

d How many votes altogether? _____

I can read and understand pictographs.

Exploring Tally Charts

1 Read the tally chart to answer the questions.

A **tally chart** is a special way to count things.

Each single tally mark stands for 1 vote. |

Each group of five tally marks stands for 5 votes. ||||/

What is your favourite toy?

Toy	Tally	Total
(ball)	||||/ ||||/ |||	
(car)	||||/ ||	
(puzzle)	||||/	

a How many children picked ? _____

b How many children picked ? _____

c How many children picked ? _____

I can read and understand tally charts.

Exploring Tally Charts (continued)

2 Read the tally chart to answer the questions.

What is your favourite time of day?

Time of Day	Tally	Total
☀️ (sunrise)	‖‖‖	
☀️ (sun)	‖‖‖ ‖‖‖		
🌙 (moon)	‖‖‖ ‖‖‖	

a) How many children picked 🌅 ?

b) How many children picked ☀️ ?

c) How many children picked 🌙 ?

d) Circle the **most popular** time of day.   🌅   ☀️   🌙

I can read and understand tally charts.

Exploring Bar Graphs

A **bar graph** is a way to show data using shaded bars.

1 Read the bar graph to answer the questions.

What is your favourite hobby?

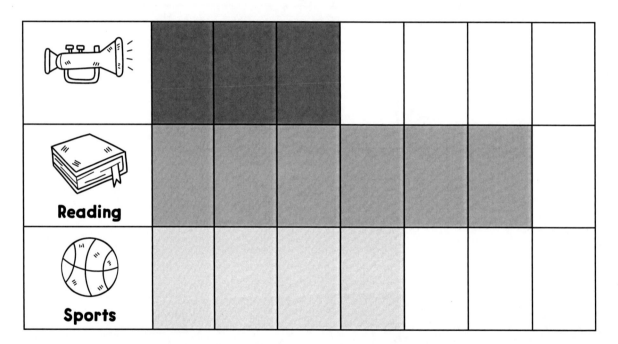

a How many votes? _____ _____ _____

b Circle the **most popular** hobby.

c Circle the **least popular** hobby.

I can read and understand bar graphs.

Exploring Bar Graphs (continued)

2 Read the bar graph to answer the questions.

What is your favourite fruit?

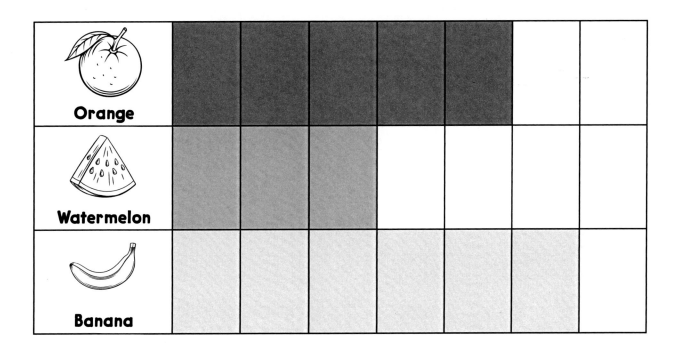

a How many votes? _____ _____ _____

b Circle the **most popular** fruit.

c Circle the **least popular** fruit.

d How many votes altogether? _____

I can read and understand bar graphs.

Collecting Data

1. Let's use the children's favourite colours to fill in the tally chart.

Colour	Tally	Total
Red		
Blue		
Purple		

I can collect data.

Collecting Data (continued)

2 Let's create and fill in a tally chart. Choose a topic.

_____

	Tally	Total

Tell one thing your tally chart shows.

_____

_____

I can collect data.

Sorting Objects

1 Look at the objects.
   Circle the correct sorting rule.

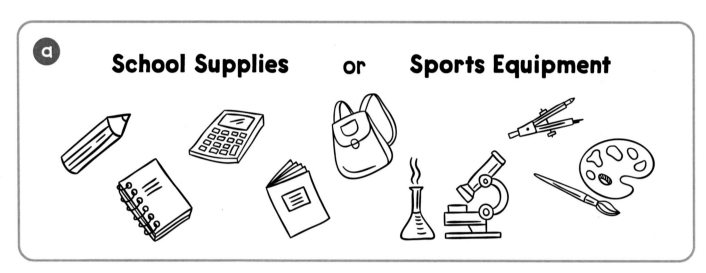

a School Supplies or Sports Equipment

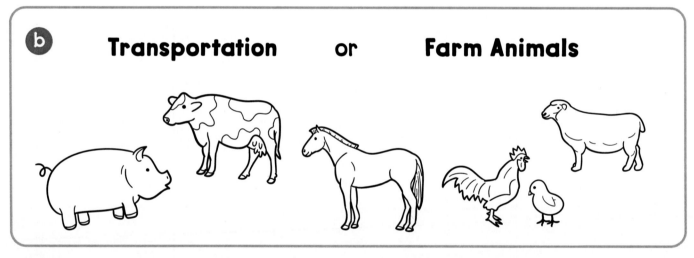

b Transportation or Farm Animals

c Vegetables or Sweets

I can identify the sorting rule.

Sorting Objects (continued)

2 Look at the objects.

Colour the objects that go in basket A (red).

Colour the objects that go in basket B (blue).

Sorting Objects (continued)

3 Look at object in the box on the left.
Circle the object on the right that is **exactly the same**.

4 Look object in the box on the left.
Circle the object on the right that is **different**.

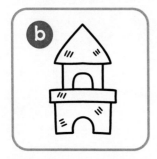   

I can identify the sorting rule.

Ordering Objects

1 Number the objects from **tallest** to **shortest**.

_____ , _____ , _____ , _____

2 Number the objects from **biggest** to **smallest**.

_____ , _____ , _____ , _____

3 Number the objects from **thinnest** to **widest**.

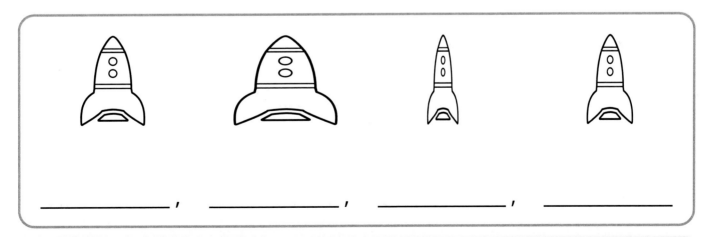

_____ , _____ , _____ , _____

I can order objects.

Show What You Know!

1 Read the bar graph to answer the question.

What is your favourite sport?

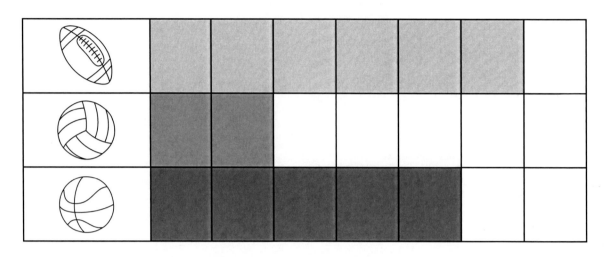

How many votes? ____   ____ 🏀 ____

2 Read the tally graph to answer the question.

Do you like winter or summer?

Season	Tally	Total
☀️	ⅢⅢ ⅢⅢ Ⅰ	
⛄	ⅢⅢ Ⅲ	

Circle the **most popular** season.

Show What You Know! (continued)

3 Number the objects from **biggest** to **smallest**.

_____ , _____ , _____ , _____

4 Look at the objects.

Colour the objects that go in basket A 🖍 red.
Colour the objects that go in basket B 🖍 blue.

Animals   Flowers

I Can Checklist:
Probability

 Use mathematical language, including the terms "impossible," "possible," and "certain," to describe the likelihood of events happening. For example:
- ***I can*** *use the terms "impossible", "possible", and "certain" to describe likelihood.*

 Use the likelihood of events happening to make predictions and informed decisions.
- ***I can*** *use the likelihood of events to make predictions.*

 Use the terms "impossible," "possible," and "certain" to describe the likelihood of events happening. For example:
- ***I can*** *say that it is impossible for the sun to rise in the west.*

 Use likelihood to make predictions and informed decisions. For example:
- ***I can*** *predict that it is possible for it to rain today, based on the dark clouds in the sky.*

 Use mathematical language to describe the likelihood of events. For example:
- ***I can*** *say that it is certain that I will get a sticker if I answer a question correctly.*

 Understand that impossible means something will never happen. For example:
- ***I can*** *understand that it is impossible for a cat to bark like a dog. I can understand that possible means something could happen but is not guaranteed.*

 Understand that certain means something is guaranteed to happen. For example:
- ***I can*** *understand that it is certain that the sun will rise every morning.*

Thinking About Likelihood

Circle if is **possible** or **impossible** for the event to happen.

1

What is the probability of you playing with your friend?

possible   impossible

2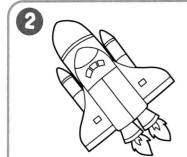

What is the probability of taking a ride on a rocket ship tomorrow?

possible   impossible

3

What is the probability of eating an ice cream cone?

possible   impossible

4

What is the probability of meeting a talking dog?

possible   impossible

I can understand likelihood.

Exploring Probability

1. Circle the best answer for each question.

a  What is the probability of seeing a tree walk?

certain   impossible

b  What is the probability of a dinosaur running through your yard?

certain   impossible

c  What is the probability of snow falling during the summer?

certain   impossible

d  What is the probability of snow falling during the winter?

certain   impossible

I can explore and understand probability.

Exploring Probability (continued)

2 Circle the best answer for each question.

 Today you will fly a plane.

possible   certain   impossible

 Today you will read a book.

possible   certain   impossible

 Today you will have a drink.

possible   certain   impossible

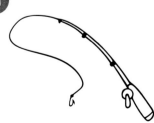 Today you will go fishing.

possible   certain   impossible

I can explore and understand probability.

Show What You Know!

1. Circle the best answer for the question.

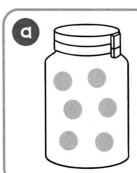

What is the probability of getting a ▪ ?

possible   certain   impossible

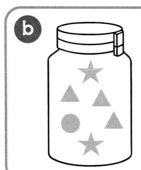

What is the probability of getting a ● ?

possible   certain   impossible

What is the probability of getting a ★ ?

possible   certain   impossible

What is the probability of getting a ★ ?

possible   certain   impossible

Show What You Know! (continued)

2 Circle the best answer for the question.

a) Today you will go to bed at the end of the day.

possible   certain   impossible

b) Today you will play at the playground.

possible   certain   impossible

c) What is the probability of an alien stopping by for dinner tonight?

possible   certain   impossible

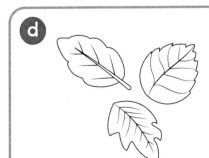

d) What is the probability of the leaves changing colour in the fall?

possible   certain   impossible

I Can Checklist:
Geometric and Spatial Reasoning

Geometric Reasoning

✓ Sort three-dimensional objects and two-dimensional shapes based on one attribute at a time. For example:
- ***I can*** *sort shapes based on their colour or size.*

✓ Identify the sorting rule being used when objects are sorted. For example:
- ***I can*** *say that objects are sorted based on their shape or number of sides.*

✓ Construct three-dimensional objects and recognize two-dimensional shapes within them. For example:
- ***I can*** *build a block tower and identify the square and rectangle shapes on the sides.*

✓ Construct and describe two-dimensional shapes and three-dimensional objects that have matching halves. For example:
- ***I can*** *create a symmetrical butterfly by folding a paper in half.*

Location and Movement

✓ Describe the relative locations of objects or people using positional language. For example:
- ***I can*** *say that the book is on the table or the ball is under the chair.*

✓ Give and follow directions for moving from one location to another. For example:
- ***I can*** *give directions to a friend for how to go from the classroom to the library.*

Colouring 2D Shapes

Colour the shapes using the colour key.

Colour the circles. — green
Colour the squares. — orange
Colour the rectangles. — blue
Colour the triangles. — red
Colour the ovals. — purple
Colour the octagons. — yellow

Exploring 2D Shapes

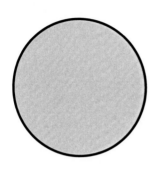

Circle the name of this shape.

    square   circle

How many sides does it have? _____

How many corners does it have? _____

Trace the shape. Draw the shape.

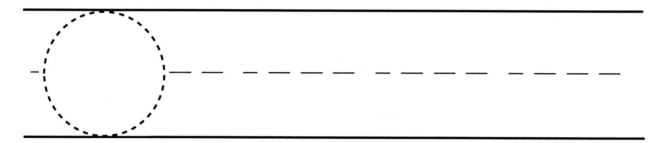

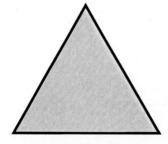

Circle the name of this shape.

    triangle   rectangle

How many sides does it have? _____

How many corners does it have? _____

Trace the shape. Draw the shape.

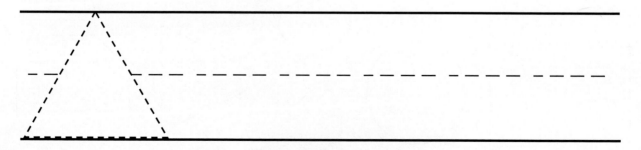

I can identify 2D shapes.

Exploring 2D Shapes (continued)

Circle the name of this shape.

hexagon   square

How many sides does it have? _____

How many corners does it have? _____

Trace the shape. Draw the shape.

Circle the name of this shape.

circle   rectangle

How many sides does it have? _____

How many corners does it have? _____

Trace the shape. Draw the shape.

I can identify 2D shapes.

Exploring 2D Shapes (continued)

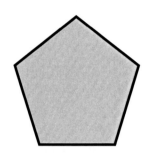

Circle the name of this shape.

circle   pentagon

How many sides does it have? _____

How many corners does it have? _____

Trace the shape. Draw the shape.

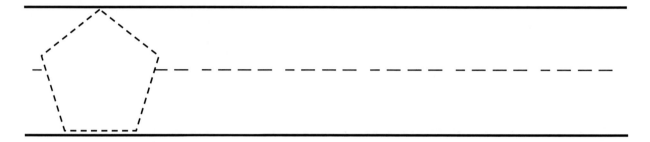

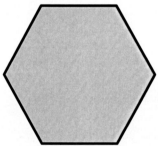

Circle the name of this shape.

triangle   hexagon

How many sides does it have? _____

How many corners does it have? _____

Trace the shape. Draw the shape.

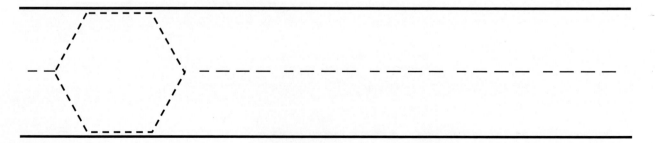

I can identify 2D shapes.

Exploring 2D Shapes (continued)

Circle the name of this shape.

octagon   **trapezoid**

How many sides does it have? _____

How many corners does it have? _____

Trace the shape. Draw the shape.

Circle the name of this shape.

rectangle   **trapezoid**

How many sides does it have? _____

How many corners does it have? _____

Trace the shape. Draw the shape.

I can identify 2D shapes.

Shapes All Around Us!

You can find shapes all around you! Look for shapes inside and outside your home!

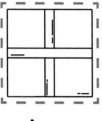
box

clock

ice cream cone

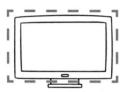
television

kite

light bulb

blocks

egg

door

wheel

pizza

road signs

I can find shapes all around me.

Take a 2D Shape Walk

No matter where you live, you can see different shapes outside. With an adult helper, make a list of things you find that are the shapes below.

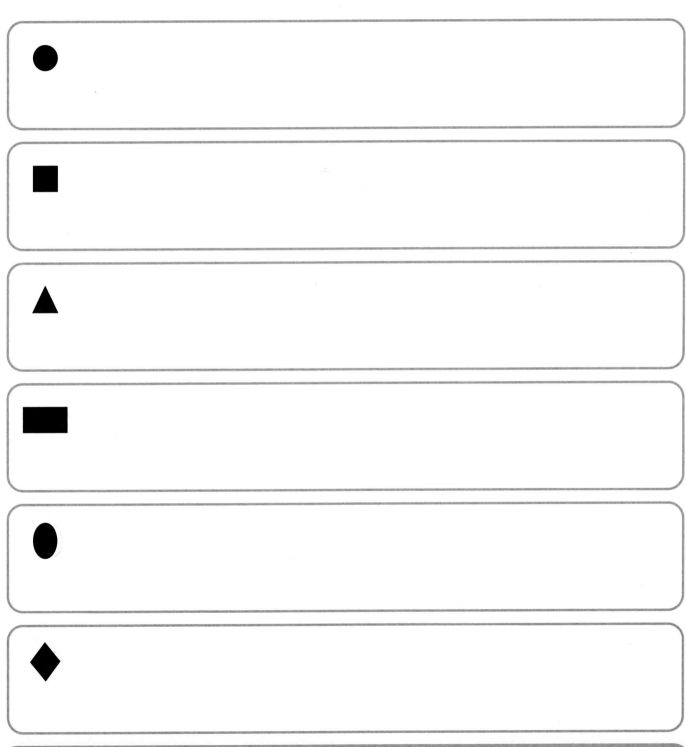

I can see different shapes outside.

Sorting 2D Shapes

Read the sorting rule. Colour the shapes that follow the rule.

1 Shapes with **four sides**.

  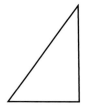 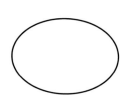

2 Shapes with **three corners**.

 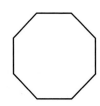  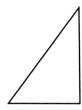

3 Shapes with more than **four sides**.

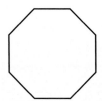 

4 Shapes with **eight corners**.

  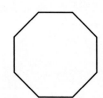 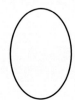

I can sort 2D shapes.

Draw the Other Half

1 Draw the other half of the shape.

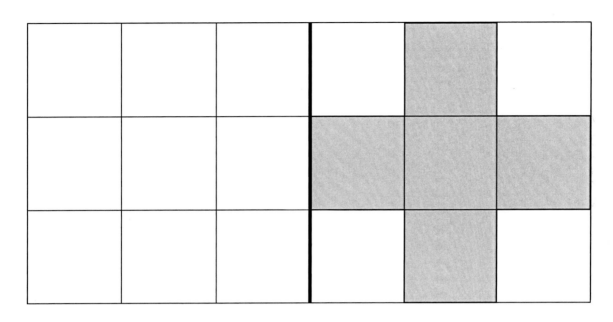

2 Draw a line from the shape on the left to its other half on the right.

a

b

c

I can draw the other half of a shape.

Exploring 3D Figures

Draw a line from the correct 3D figure to the object. Think about how these everyday objects and three-dimensional figures are similar or different. Explain your thinking to a friend.

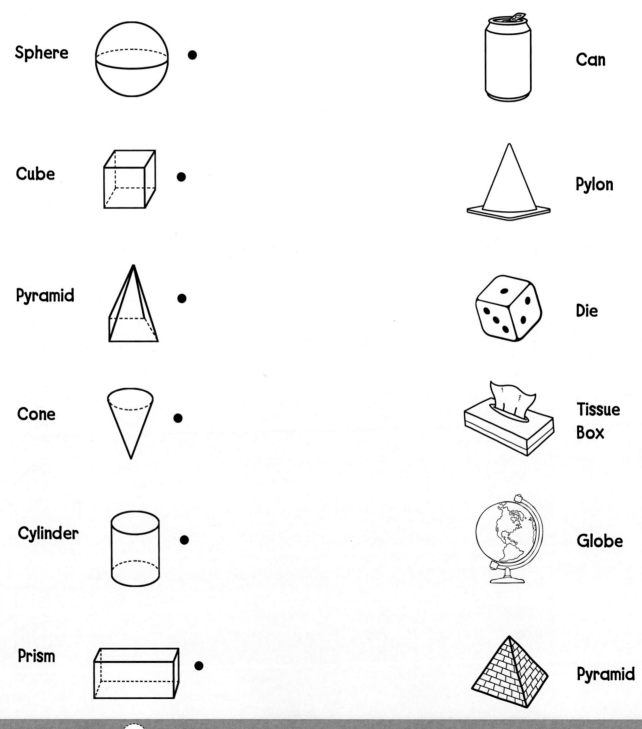

I can explore and identify 3D shapes.

Sorting 3D Figures

Follow the sorting rule.

1 Circle the 3D figures that **can roll**.

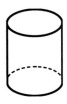    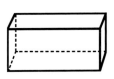

2 Circle the 3D figures that **cannot roll**.

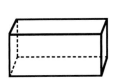

3 Circle the 3D figures that **can stack** on each other.

  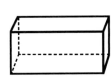

4 Circle the 3D figures that **cannot stack** on each other.

 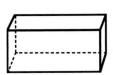 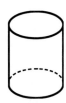

I can sort 3D figures.

Where Is the Hamster?

Draw a line from the hamster to the correct positional word.

   • behind

   • between

   • in front of

   • above

   • inside

I can explore and understand positional words.

Where Is the Dog?

Draw a line from the dog to the correct positional word.

 • • under

 • • near

 • • beside

 • • on

I can explore and understand positional words.

Positional Words: Follow the Instructions

1 Colour the astronaut **behind** the rocket ship (red).
Colour the stars **on the left** of the rocket ship (yellow).
Colour the stars **below** the Earth (blue).
Colour the asteroid **next to** the astronaut (green).
Colour the asteroid **under** the rocket ship (purple).

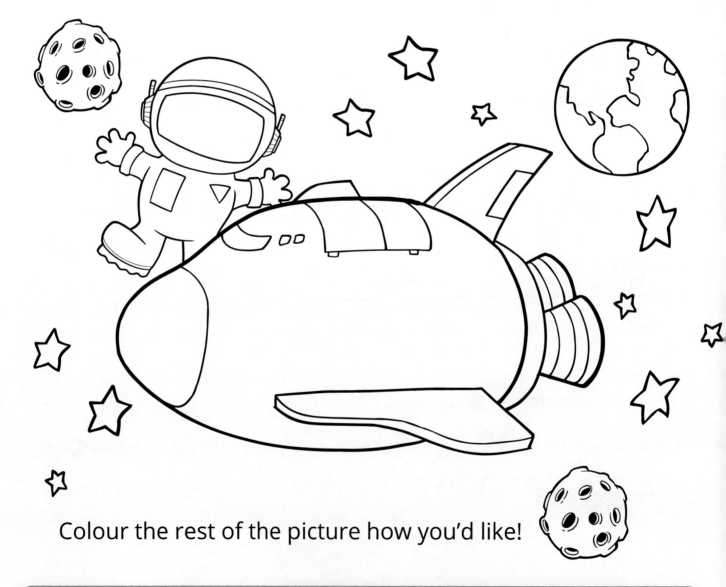

Colour the rest of the picture how you'd like!

I can explore and understand positional words.

Exploring Location and Movement

Follow the instructions.

1 Circle the animal that is 2 spaces right and 1 space up from the .

2 Circle the animal that is 1 space left and 2 spaces down from the .

3 Draw a ◯ 2 spaces to the right of the .

4 Draw a △ 3 spaces to the left of the .

I can explore location and movement.

Show What You Know!

1. Follow the instructions for the 2D shape.

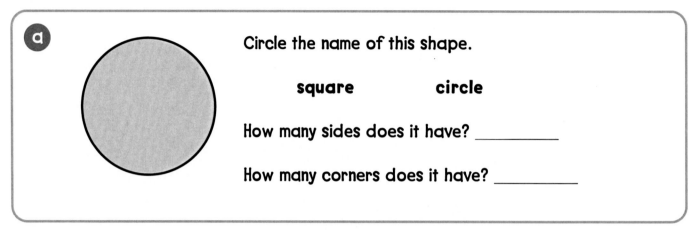

a Circle the name of this shape.

square   circle

How many sides does it have? _____

How many corners does it have? _____

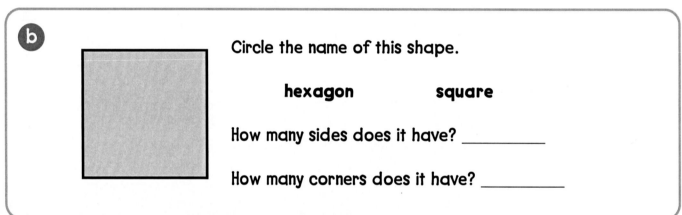

b Circle the name of this shape.

hexagon   square

How many sides does it have? _____

How many corners does it have? _____

2. Draw a line from the 3D figure to the object with the same shape.

Show What You Know! (continued)

3 Colour the shapes with four sides.

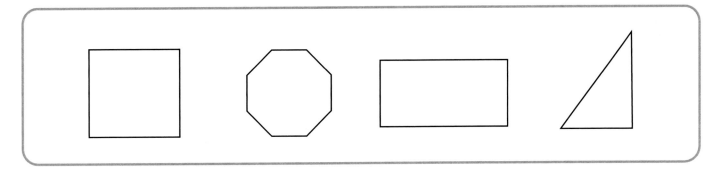

4 Circle the 3D figures that can roll.

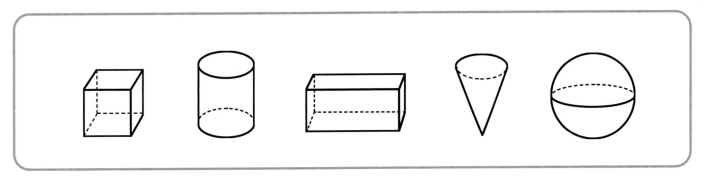

5 Draw a line from the animal to the correct positional word.

• under

• above

I Can Checklist: Measurement

Attributes

- ✓ Compare two objects using measurable attributes (like length, mass, or capacity). For example:
  - *I can* pick two pencils and tell which one is longer.

- ✓ Explain why the choice of unit (such as paperclips, cubes, etc.) can affect the measurement of length. For example:
  - *I can* understand that if I use larger blocks to measure my book, I will need fewer blocks than if I use smaller cubes.

- ✓ Estimate, measure, and record lengths, heights, and distances using non-standard units. For example:
  - *I can* guess how many hand spans long the table is, then measure it with my hand and record the result.

- ✓ Estimate, measure, and record the capacity of a container using non-standard units. For example:
  - *I can* guess how many cupfuls of water my toy bucket can hold, then fill it up and record the result.

- ✓ Estimate, measure, and record the mass of an object using non-standard units. For example:
  - *I can* guess how many marbles weigh the same as my toy car, then balance them on a scale and record the result.

- ✓ Choose a personal referent for a metre to help with estimations of measurements in metres. For example:
  - *I can* understand that my height is close to a metre and use that to guess how many 'mes' tall a tree is.

Exploring Measurement: Non-Standard Units

1 Count the blocks to measure the objects.
This key is **4 blocks long**.

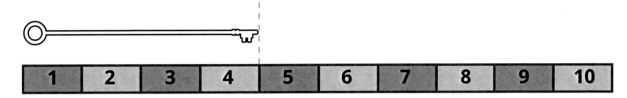

_____ blocks

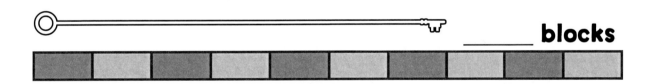

_____ blocks

2 Draw a line that is **4 blocks long**.

I can explore measurement with non-standard units.

Exploring Measurement:
Non-Standard Units (continued)

3 Count the blocks to measure the object.

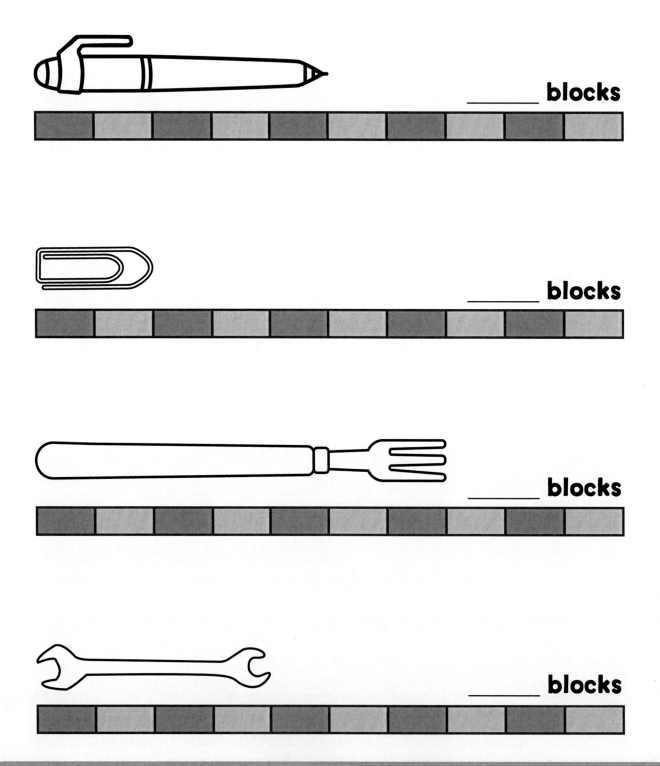

Exploring Measurement:
Centimetres

1 Measure the objects below.

a 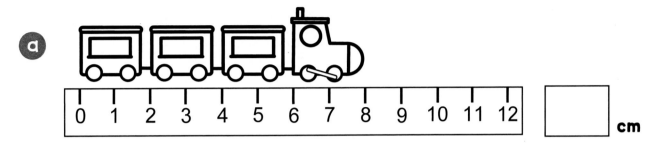 cm

b 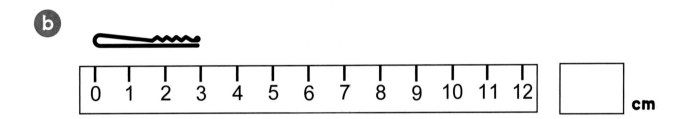 cm

c 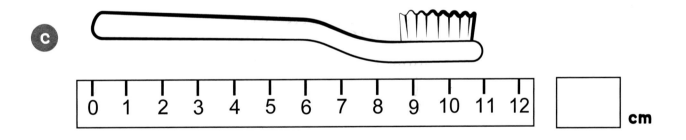 cm

d 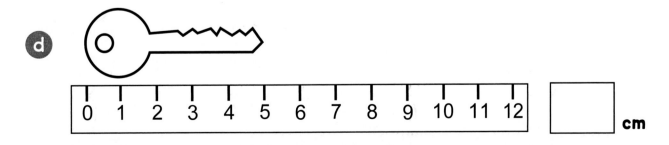 cm

I can explore measurement with centimetres.

Exploring Measurement:
Centimetres (continued)

2 Measure the object below.

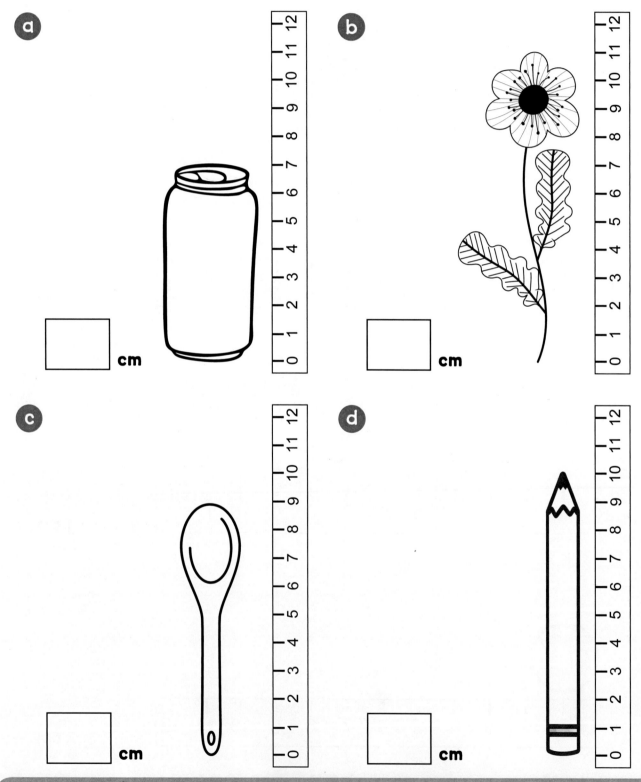

Comparing Length

Length measures how long an object is.

1 Circle the **longer** object.

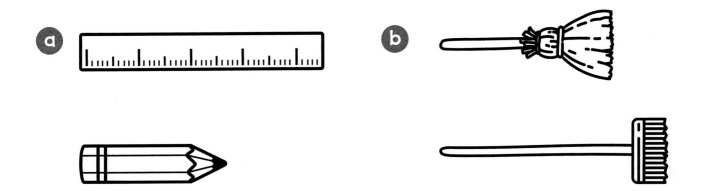

2 Number the objects from **shortest** to **longest**.

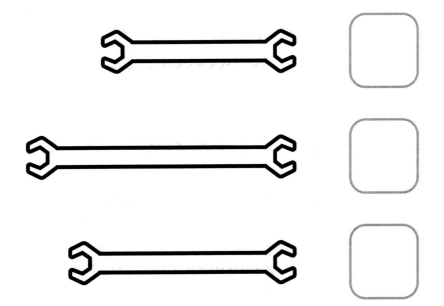

I can compare length.

Comparing Height

Height tells how tall someone or something is.

1 Circle the **taller** object.

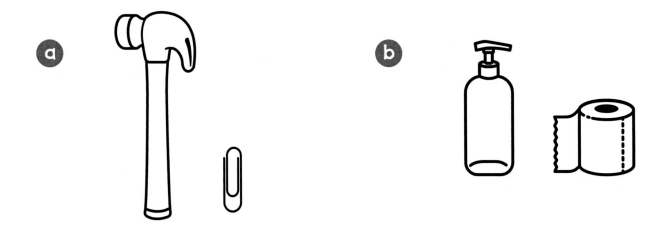

2 Number the buildings from **shortest** to **tallest**.

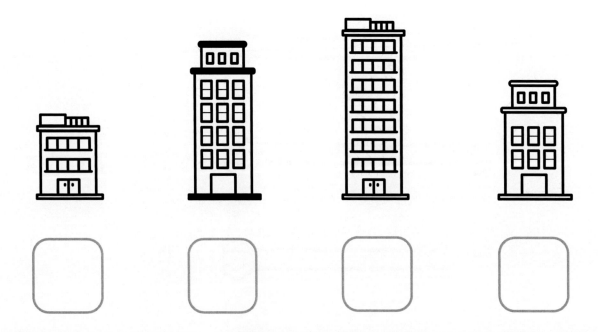

I can compare height.

Comparing Capacity

Capacity is the amount of something a container or object can hold.

1 Which container **holds more**?

2 How much do the containers hold? Number the containers from the **least** to the **most**.

I can compare capacity.

Comparing Mass

Mass measures how much something weighs.

1 What is the **mass** of the object? Count the blocks to find out.

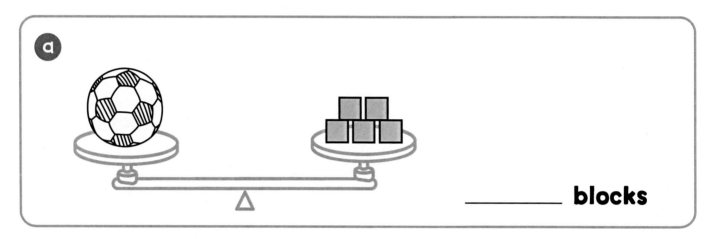

_____ **blocks**

_____ **blocks**

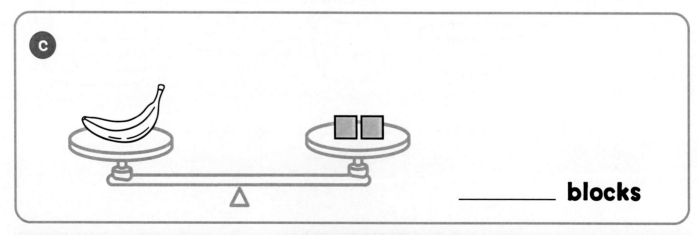

_____ **blocks**

I can compare mass.

Comparing Mass (continued)

2 Circle the **heavier** object.

a   b

3 Draw something **lighter** on the scale.

I can compare mass.

Show What You Know!

1 Follow the instructions for the question.

a Circle the **shorter** object.

b Circle the **lighter** object.

c Circle the **heavier** object.

d Circle the **longer** object.

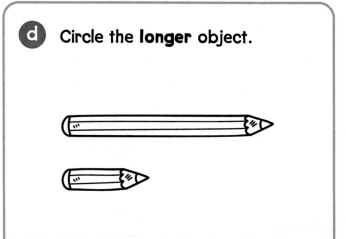

e Circle the **shorter** object.

f Circle the **taller** object.

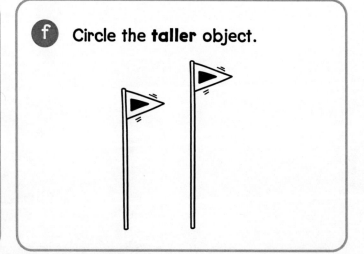

Show What You Know! (continued)

2 Count the blocks to measure the object.

_____ **blocks**

3 Measure the object below.

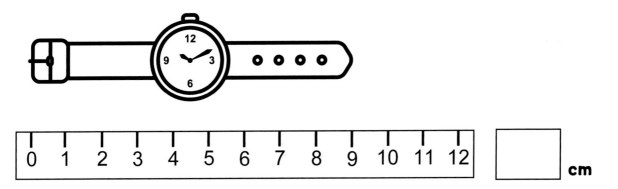

☐ **cm**

4 What is the **mass** of the object? Count the blocks to find out.

_____ blocks.

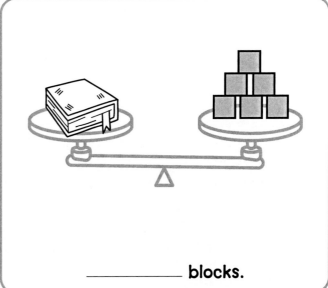

_____ blocks.

I Can Checklist:
Time

Time

 Tell time to the half hour using analog and digital clocks. For example:
- ***I can** look at an analog clock and say it is 1:30, and do the same with a digital clock.*

 Relate the concept of time (in hours and half hours) to the activities in my daily schedule. For example:
- ***I can** understand that I start school at 9 o'clock and have lunch at 12 o'clock.*

 Name the days of the week and the months of the year in order. For example:
- ***I can** recite all seven days of the week and all twelve months of the year, in order.*

 Read dates using a calendar. For example:
- ***I can** point out today's date on the calendar.*

Parent Tips:

Make It Fun:
Turn learning into a game! Use flashcards or board games that involve time and dates. There are also plenty of educational apps available that make learning time and dates fun and interactive.

Use a Real Clock:
While digital clocks are everywhere, it's best to start with an analog clock when teaching time. The visual representation of the hands on the clock makes the concept of time more tangible.

Daily Time Checks:
Ask your child what time it is at various points throughout the day. This can help them get used to the idea of time passing.

Complete a Clock Face

Write the numbers on the clock. Draw the minute hand to 12. Draw the hour hand to 6.

Telling Time to the Hour

A clock has an hour hand.
The hour hand is short. It shows the hour.

You can write the time in two ways.
It is **5 o'clock** or **5:00**.

1 Draw a line between the times that are the same.

4 o'clock •   • 2:00

12:00 •   • 4:00

2 o'clock •   • 7 o'clock

7:00 •   • 12 o'clock

2 Write the time in two ways.

a    b

_____ o'clock or   _____ o'clock or

_____:00   _____:00

I can tell time to the hour.

Telling Time to the Hour (continued)

3 Tell the time to the hour. Highlight the hour hand in blue.

a

_____ o'clock

b

_____ o'clock

c

_____ o'clock

d

_____ o'clock

e

_____ o'clock

f

_____ o'clock

I can tell time to the hour.

Telling Time to the Half Hour

A clock has an hour hand. The hour hand is short. It shows the hour.

It is **3 o'clock** or **3:00**. There are 60 minutes in an hour.

A clock has a minute hand. The minute hand is long. It shows the minutes after the hour.

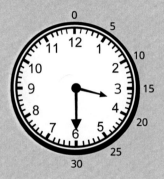

Count by 5s.
It is 30 minutes after 3 o'clock.
It is **half past 3** or **3:30**.

1 What time is it? Write the time in words.

a

b
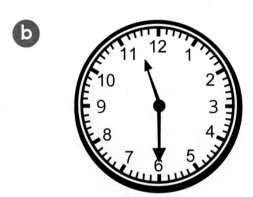

_____   _____

I can tell time to the half hour.

Telling Time to the Half Hour (continued)

2 Draw a line from the clock to the matching time.

 •   • 5:30

 •   • 10:30

 •   • 4:30

 •   • 9:30

I can tell time to the half hour.

Exploring Digital Clocks

1 Draw a line from the analog clock to the digital clock that has the matching time.

   •   •

   •   •

2 Read the instructions. Circle the correct clock.

a Circle the clock showing **12 o'clock**.

      12:00   12:30

b Circle the clock showing **half past 2**.

      2:00   2:30

I can read digital clocks.

A.M. and P.M.

10 A.M.

A.M. means before noon

10 P.M.

P.M. means after noon

Example: School starts at 9 o'clock in the morning. What time is it?

Read the instructions. Circle the correct clock.

1 Sam plays with her dog in the **morning**. What time is it?

7:00 A.M.   7:00 P.M.

2 Annie eats an apple for an **afternoon** snack. What time is it?

3:00 A.M.   3:00 P.M.

I can understand A.M and P.M.

Months of the Year

1 List the months of the year in the correct order.

Word Bank	
August	1 _____
January	2 _____
May	3 _____
September	4 _____
June	5 _____
February	6 _____
July	7 _____
October	8 _____
March	9 _____
November	10 _____
April	11 _____
December	12 _____

I can identfiy the months of the year.

Reading a Calendar

November						
Sunday	Monday	Tuesday	Wednesday	Thursday	Friday	Saturday
	1 Music	2	3 Gym	4	5	6
7	8 Music	9 Pizza Day	10 Gym	11	12	13
14	15 Music	16	17 Gym	18	19 Field Trip	20
21	22 Music	23 School Fair	24 Gym	25	26	27
28	29 Music	30				

1 Name the month for this calendar.

---

2 What day of the week is the first day of the month?

---

3 What day of the week is the last day of the month?

---

4 What is the date of the field trip?

I can read a calendar.

Show What You Know!

1 Draw a line between the times that are the same.

8 o'clock •					• 6:00

3:00 •					• 8:00

6 o'clock •					• 3 o'clock

1:00 •					• 5:00

5 o'clock •					• 12 o'clock

12:00 •					• 1 o'clock

2 Write the time in two ways.

_____ o'clock or			_____ o'clock or

_____:00				_____:00

Show What You Know! (continued)

3 Tell the time to the half hour. Highlight the hour hand in blue. Highlight the minute hand in red.

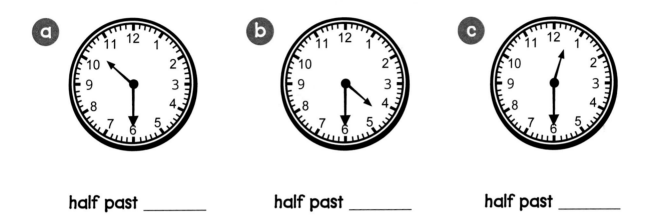

a) half past _____

b) half past _____

c) half past _____

4 Read the instructions. Circle the correct clock.

a) Circle the clock showing **1 o'clock**.

b) Trevor finishes soccer practice in the **afternoon**. What time is it?

Congratulations!
Great Work!

Answers

Number Word Search p. 6

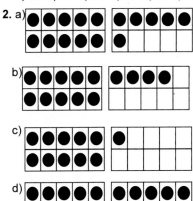

Using Ten Frames to Count to 10 pp. 8–9

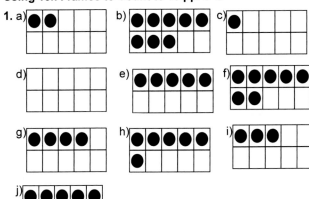

2. a) 10 b) 5 c) 4 d) 3 e) 7 f) 8 g) 9 h) 6 i) 2 j) 0

Using Ten Frames to Count to 20 pp. 10–11

1. a) 13 b) 19 c) 17 d) 20 e) 12 f) 15

2. a), b), c), d) [ten frame diagrams]

Tens and Ones to 50 pp. 12–13

1. a) 33 b) 25   2. a) 44 b) 36 c) 18

Counting Forward to 100 p. 14

1. a) 16, 17, 18, 19, 20 b) 11, 12, 13, 14, 15
2. a) 25, 30, 35, 40, 45 b) 20, 25, 30, 35, 40
3. a) 20, 30, 40, 50, 60 b) 60, 70, 80, 90, 100

Counting Backward from 100 p. 15

1. a) 7, 6, 5, 4, 3 b) 9, 8, 7, 6, 5
2. a) 45, 40, 35, 30, 25 b) 25, 20, 15, 10, 5
3. a) 50, 40, 30, 20, 10 b) 70, 60, 50, 40, 30

Ordering Numbers pp. 18–19

1.
1	2	3	**4**	5	6	7	8	9	10
11	12	**13**	14	15	16	17	18	**19**	20
21	22	23	24	25	26	**27**	28	29	30
31	**32**	33	34	35	36	37	**38**	39	40
41	42	43	44	45	**46**	47	48	49	**50**

2. a) 10 b) 50 c) 21
3. a) 34, 40, 50 b) 20, 22, 48 c) 14, 17, 21   4. a) 49, 34, 1
   b) 52, 33, 27 c) 36, 20, 17

Comparing Numbers from 1 to 10 p. 20

1. a) < b) < c) > d) =

Comparing Numbers to 50 p. 21

1. a) > b) < c) > d) =

Estimating and Counting pp. 22–23

Estimations will vary.

1. a) 7 b) 5   2. a) 6 b) 8 c) 7

Show What you Know! pp. 24–25

1. a) 20 b) 17   2. 23   3. Estimations will vary. 8
4. a) < b) =   5. a) 50, 44, 22 b) 52, 49, 48   6. Ten

Exploring Fair Sharing pp. 27–29

1. Each cupcake should be circled using a different colour.
2. a) 3 b) 2   3. 3

More Fair Sharing pp. 30–31

1. a) YES b) NO   2. a) NO b) YES

Exploring More Fair Sharing pp. 32–33

1. a) Each child gets 1. 1 left over   b) Each child gets 1. 1 left over
   c) Each child gets 1. 2 left over
2. a) Each child gets 2. 1 left over   b) Each child gets 3. 1 left over
   c) Each child gets 1. 2 left over

Exploring Equal Parts: Halves pp. 34–35

1.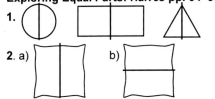

2. a)   b)

3. a) 4 b) 2 c) 5

Exploring Equal Parts: Fourths pp. 36–37

1.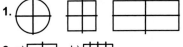

2. a)   b)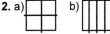

3. a) 1 b) 2 c) 3

Exploring Equal Parts: Halves and Fourths pp. 38–39

Ensure students colour the correct fractions.

© Chalkboard Publishing Inc

Answers

Colour the Fractions p. 40

1.   2.

3.   4.

5.   6.

Working with Fractions p. 41

1. Right shape 2. Left shape
3. 1/2 4. 1/4 5.   6. Left shape

Ordering Fractions p. 42

1. 1 3 4 2   2. 2 1 3   3. 3 4 1 6 2 5

Comparing Fractions p. 43

1. = 2. > 3. > 4. < 5. < 6. =

Show What You Know! pp. 44–45

1. a) 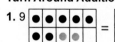 b)

2. a) < b) = c) > d) <   3. YES   4. 5 2 4 1 3 6

Plus Zero Addition Strategy p. 47

1. 10 2. 5 3. 6 4. 3 5. 18 6. 15

Count On Addition Strategy p. 48

1. 7 2. 5, 6 3. 13 4. 19, 20

Using a Number Line to Add p. 49

1. 7 2. 10 3. 9

Turn Around Addition Strategy p. 50

1. 9 =

2. 9 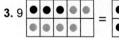 =

3. 9 (ten frames)

Counting Doubles Addition Strategy p. 51

1. 2 + 2 = 4   2. 4 + 4 = 8   3. 7 + 7 = 14   4. 5 + 5 = 10

Doubles Plus One Addition Strategy p. 52

1. 6, 7   2. 8, 9   3. 10, 11   4. 12, 13

Draw a Picture Addition Strategy p. 53

1. 11   2. 13   3. 12   4. 16

Addition Word Problems pp. 54–56

1. a) 3 + 1 = 4   b) 6 + 4 = 10   c) 6 + 9 = 15
2. a) 5 + 9 = 14   b) 10 + 5 = 15   c) 6 + 3 = 9
3. a) 9 + 2 = 11   b) 2 + 5 = 7   c) 7 + 2 = 9

Adding Ten More p. 57

1. a) 30 + 10 = 40   b) 50 + 10 = 60
2. a) 50 b) 30 c) 70 d) 90 e) 20 f) 80

Addition: Sums to 20 p. 59

1. a) 13 b) 9 c) 12 d) 16 e) 11 f) 20 g) 12 h) 13
   i) 17 j) 17 k) 10 l) 11

Minus Zero Subtraction Strategy p. 60

1. 5   2. 6   3. 7   4. 13   5. 9   6. 20

Counting Back Subtraction Strategy p. 61

1. 7   2. 5, 4   3. 6   4. 8, 7

Using a Number Line to Subtract p. 62

1. 4   2. 3   3. 6   4. 8

A Number Minus Itself p. 63

1. 0   2. 0   3. 0   4. 0   5. 0   6. 0   7. 0   8. 0

Doubles Subtraction Strategy p. 64

1. 6 − 3 = 3   2. 10 − 5 = 5   3. 2 − 1 = 1   4. 4 − 2 = 2

Draw a Picture Subtraction Strategy p. 65

1. 5   2. 7   3. 8   4. 15

Subtraction Word Problems pp. 66–68

1. a) 10 − 6 = 4   b) 20 − 10 = 10   c) 5 − 4 = 1
2. a) 17 − 12 = 5   b) 18 − 7 = 11   c) 16 − 7 = 9
3. a) 12 − 6 = 6   b) 10 − 5 = 5   c) 5 − 4 = 1

Ten More, Ten Less p. 69

10 Less Column: 20, 80, 40, 10, 50, 70
10 More Column: 40, 100, 60, 30, 70, 90

Subtracting Numbers from 0 to 20 p. 71

1. 2   2. 8   3. 10   4. 8   5. 14   6. 0   7. 10   8. 13   9. 13   10. 16
11. 3   12. 8

Show What You Know — Sums from 0 to 10 p. 72

Top: 4, 2, 6, 5, 7, 3, 6; Middle: 7, 8, 1, 2, 9, 6, 8; Bottom: 3, 5, 9, 10, 10, 0

Show What You Know — Sums from 11 to 20 p. 72

Top: 20, 16, 20, 15, 19, 14, 17; Middle: 19, 16, 17, 12, 18, 15, 13; Bottom: 13, 15, 11, 20, 14, 12

Show What You Know — Subtraction from 0 to 10 p. 73

Top: 5, 2, 6, 3, 8, 4, 2; Middle: 2, 2, 3, 0, 0, 4, 2; Bottom: 2, 0, 4, 2, 1, 5

Show What You Know — Subtraction from 11 to 20 p. 73

Top: 7, 7, 10, 7, 10, 4, 10; Middle: 16, 10, 4, 12, 11, 11, 9; Bottom: 6, 8, 12, 13, 15, 14

Exploring Equal Groups pp. 74–77

1. a) 2, 10   b) 4   c) 2, 6   2. a) 2   b) 8   c) 5, 10

Answers

3. a) 2 b) 8 c) 5   **4.** a) 4, 12 b) 3 c) 5

Patterns Are All Around Us! p. 79

Answers will vary. Sample answers: Inside: measures on a ruler, patterns on clothing, buttons on clothing, patterns on upholstery, hole patterns on shoes; In nature: layers on a pine cone, windows on a building, petals on a flower

Patterns in Everyday Life p. 81

1. a)   b)

2. Answers will vary.

Extend the Pattern pp. 82–83

1. a) A B A   b)   c)   d)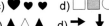

2. a)    b) ■■□   c) ▲△▲   d) →↓→   e) A B C

Growing Patterns p. 84

1. a) 7, 8, 9. Start at 1 then add 1 each time.
   b) 12, 14, 16. Start at 2 then add 2 each time

2.

Shrinking Patterns p. 85

1. a) 5, 4, 3   b) 16, 15, 14   c) 10, 9, 8

2.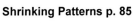

How Does the Pattern Attribute Change? pp. 86–87

1. a) colour   b) position   c) size

2. a) size   b) position   c) size   d) colour

Describing the Pattern Rule p. 88

1. AABAABA   2. ABABABA

Identifying and Describing Patterns p. 89

1. AAB   2. ABC   3. ABBC   4. AABB

Complete the Pattern pp. 90–91

1. a) ☐   b) 35 30   c) ↑ ↑
   d) ○ ○

2. a) △   b) ■ ■   c) ♥
   d) 8 14

Create a Pattern p. 92

1. – 4. Answers will vary.

Create a Pattern Rule p. 93

1. and 2. Answers will vary.

More and Fewer Objects pp. 96–97

1. a) 4 circles should be drawn.
   b) 3 triangles should be drawn.
   c) 2 triangles should be drawn.

2. a) 3 triangles should be drawn.
   b) 4 squares should be drawn.
   c) 2 circles should be drawn.

Show What You Know! pp. 98–99

1. △ ○   2. ↓ ↑   3. 10 9 8
   ▲ ▲
4. ▲ ▲   5. a) ABC   b) AAB   c) ABBC   6. colour
   ▲ ▲

Understanding Things That Stay the Same and Things That Can Change p. 101

Orange items: 24 hours in a day, 12 months in a year number of cents in a dollar.

Green items: number of hours spent playing, number of cents to buy something, number of months until a special event.

Balance It! pp. 102–103

1. b) 1   c) 4   d) 3   2. a) 2   b) 2   c) 4   d) 3

Making Addition Sentences pp. 104–105

1. Answers may vary. Sample answers shown.
   a) 3 + 3, 2 + 4, 5 + 1   b) 5 + 5, 7 + 3, 6 + 4
   c) 3 + 1, 2 + 2, 1 + 3

2. Answers may vary. Sample answers shown.
   a) 6 + 2, 4 + 4, 1 + 7   b) 5 + 4, 7 + 2, 6 + 3
   c) 4 + 1, 3 + 2, 1 + 4

Making Subtraction Sentences pp. 106–107

1. Answers may vary. Sample answers shown.
   a) 4 − 2 = 2, 4 − 1 = 3, 4 − 3 = 1
   b) 8 − 4 = 4, 8 − 5 = 3, 8 − 6 = 2, 8 − 7 = 1
   c) 10 − 5 = 5, 10 − 7 = 3, 10 − 6 = 4, 10 − 8 = 2

2. Answers may vary. Sample answers shown.
   a) 7 − 4 = 3, 7 − 3 = 4, 7 − 5 = 2, 7 − 6 = 1
   b) 5 − 3 = 2, 5 − 2 = 3, 5 − 4 = 1
   c) 3 − 2 = 1, 3 − 1 = 2, 3 − 3 = 0

Number Fact Families pp. 108–109

1. a) 3 + 5 = 8, 5 + 3 = 8, 8 − 5 = 3, 8 − 3 = 5
   b) 4 + 3 = 7, 3 + 4 = 7, 7 − 4 = 3, 7 − 3 = 4

2. a) 7 + 2 = 9, 2 + 7 = 9, 9 − 7 = 2, 9 − 2 = 7
   b) 1 + 3 = 4, 3 + 1 = 4, 4 − 1 = 3, 4 − 3 = 1

Is It Equal? pp. 110–111

1. b) ≠ 5 + 1 = 6, 2 + 2 = 4   c) ≠ 3 + 6 = 9, 5 + 1 = 6
   d) ✓ 8 + 2 = 10, 7 + 3 = 10

2. a) ✓ 8 − 2 = 6, 9 − 3 = 6   b) ≠ 10 − 5 = 5, 4 − 3 = 1
   c) ≠ 6 − 4 = 2, 3 − 0 = 3   d) ✓ 10 − 6 = 4, 7 − 3 = 4

Show What You Know! pp. 112–113

1. a) 2   b) 3

2. a) ≠ 5 + 4 = 9, 7 + 1 = 8   b) ✓ 8 − 4 = 4, 6 − 2 = 4

3. 4 + 4 = 8, 6 + 2 = 8, 7 + 1 = 8

4. 6 + 4 = 10, 4 + 6 = 10, 10 − 4 = 6, 10 − 6 = 4

© Chalkboard Publishing Inc

Answers

Coding in Everyday Life p. 115
Items that should be coloured: car, video game, microwave, dishwasher, traffic light

Read the Code pp. 116–117
1. a)   b) 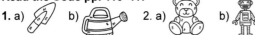   2. a)   b)

Exploring Writing Code pp. 118–119
1. a) Answers will vary. Sample answer: ↓→→→→→
   b) Answers will vary. Sample answer: ←←←↑
2. a) Answers will vary. Sample answer: ↑→→→→
   b) Answers will vary. Sample answer: ↑←←←←

Alter the Code pp. 120–121
2. Answers will vary. Sample answer: ←←←↑
4. Answers will vary. Sample answer: ↓↓→→→

Show What You Know! pp. 122–123
1. Items that should be coloured: computer, digital clock
2. a) Answers will vary. Sample answer: ←↓↓↓
   b) Answers will vary. Sample answer: ↑↑↑↑
4. Answers will vary. Sample answer: ↓↓→→→

Matching Canadian Coins p. 127

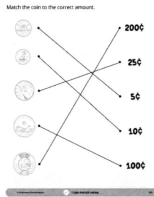

Counting Nickels p. 128
1. 5, 10, 15 = 15¢   2. 5, 10, 15, 20, 25, 30 = 30¢
3. 5, 10, 15, 20, 25 = 25¢   4. 5, 10, 15, 20, 25, 30, 35 = 35¢

Counting Dimes p. 129
1. 10, 20, 30 = 30¢   2. 10, 20, 30, 40 = 40¢
3. 10, 20, 30, 40, 50, 60 = 60¢
4. 10, 20, 30, 40, 50, 60, 70 = 70¢

Comparing Money p. 130
1. 20¢ > 10¢   2. 15¢ = 15¢   3. 10¢ < 15¢

Comparing Coin Sizes p. 131
1. Toonie should be circled.   2. Dime should be circled.
3. 1, 4, 2, 3

Comparing Bill Values p. 134
1. $5 < $10   2. $50 > $20   3. $5 < $50

Order the Coins and Bills p. 135
1. 3, 1, 2   2. 2, 1, 3   3. 1, 3, 2

Show What You Know! pp. 136–137
1. 25¢   2. 40¢   3. 50¢   4. $3
5. a) $2.05 > $1.35   b) $25 > $20

Exploring Pictographs pp. 140–141
1. a) 6, 4, 2   b)   c)
2. a) 3, 5, 6   b)   c)   d) 14

Exploring Tally Charts pp. 142–143
1. a) 13   b) 7   c) 5   2. a) 5   b) 11   c) 8   d) ☀

Exploring Bar Graphs pp. 144–145
1. a) 3, 6, 4   b)   c)
2. a) 5, 3, 6   b) 🍌   c) 🍉   d) 14

Collecting Data pp. 146–147
1. Red: || 2   Blue: |||| 4   Purple: || 2
2. Answers will vary.

Sorting Objects pp. 148–149
1. a) School supplies   b) Farm animals   c) Sweets
2. Red objects/basket A: pineapple, orange wedge, apple, strawberry, plum, banana; Blue objects/basket B: volleyball, beach ball, basketball, baseball, soccer ball, football

Sorting Objects p. 150
1. a)   b)   2. a)   b)

Ordering Objects p. 151
1. 3, 2, 4, 1   2. 2, 4, 1, 3   3. 3, 4, 1, 2

Show What You Know! pp. 152–153
1. 6, 2, 5   2. 11, 8    3. 2, 3, 4, 1
4. Red objects/basket A: fox, bird, dog, rabbit, squirrel, cat; Blue objects/basket B: Ensure that 4 flowers are coloured blue.

Thinking About Likelihood p. 155
1. possible   2. impossible   3. possible   4. impossible

Exploring Probability pp. 156–157
1. a) impossible   b) impossible   c) impossible   d) certain
2. Answers may vary, sample answers shown.
   a) impossible   b) certain   c) certain   d) possible

Show What You Know! pp. 158–159
1. a) impossible   b) possible   c) certain   d) possible
2. Answers may vary, sample answers shown.
   a) certain   b) certain   c) impossible   d) certain

Exploring 2D Shapes pp. 162–165
Circle, 0 sides, 0 corners   Triangle, 3 sides, 3 corners

Answers

Square, 4 sides, 4 corners
Pentagon, 5 sides, 5 corners
Octagon, 8 sides, 8 corners

Rectangle, 4 sides, 4 corners
Hexagon, 6 sides, 6 corners
Trapezoid, 4 sides, 4 corners

Sorting 2D Shapes p. 168

1. Square should be coloured
2. Triangle should be coloured
3. Octagon should be coloured
4. Octagon should be coloured

Draw the Other Half p. 169

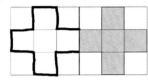
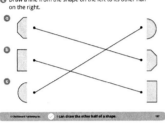

Exploring 3D Figures p. 170

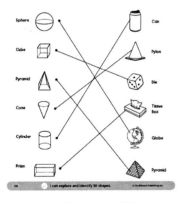

Sorting 3D Figures p. 171

1. Cylinder, sphere
2. prism, pyramid, cube
3. Cube, cylinder, prism
4. Cone, sphere

Where Is the Hamster? p. 172

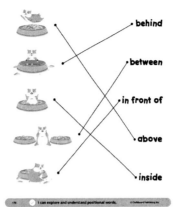

Where Is the Dog? p. 173

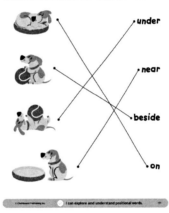

Exploring Location and Movement p. 175

1. (hamster)   2. (bird)
3. and 4.

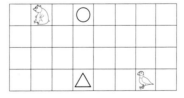

Show What You Know! pp. 176–177

1. a) Circle, 0 sides, 0 corners   b) Square, 4 sides, 4 corners
2. Sphere → Earth   Cone → Ice-cream cone
3. Square and rectangle should be coloured
4. Cylinder, cone, and sphere should be coloured
5. Hamster → above   Dog → under

Exploring Measurement: Non-Standard Units pp. 179–180

1. 3 blocks, 7 blocks

© Chalkboard Publishing Inc

Answers

2. _____

3. 5 blocks, 2 blocks, 7 blocks, 4 blocks

Exploring Measurement: Centimetres pp. 181–182

1. a) 8 cm   b) 3 cm   c) 11 cm   d) 5 cm

2. a) 7 cm   b) 11 cm   c) 9 cm   d) 10 cm

Comparing Length p. 183

1. a) Ruler should be circled   b) Bottom broom should be circled

2. 1, 3, 2

Comparing Height p. 184

1. a) Hammer should be circled   b) Soap pump should be circled

2. 1, 3, 4, 2

Comparing Capacity p. 185

1. a) Jug should be circled   b) Milk carton should be circled

2. 4, 2, 3, 1, 5

Comparing Mass pp. 186–187

1. a) 5 blocks   b) 9 blocks   c) 2 blocks

2. a) Bird should be circled   b) Cake should be circled

3. Ensure students draw something reasonably smaller.

Show What You Know! pp. 188–189

1. a) Left can   b) Leaf   c) Car   d) Top pencil   e) Top guitar
   f) Right flag

2. 3 blocks   **3.** 9 cm   **4.** a) 3 blocks   b) 6 blocks

Complete a Clock Face p. 191

Telling Time to the Hour p. 192

1.
4 o'clock — 2:00
12:00 — 4:00
2 o'clock — 7 o'clock
7:00 — 12 o'clock

2. a) 8:00   b) 5:00

3. a) 4:00   b) 10:00   c) 9:00   d) 5:00   e) 6:00   f) 7:00

Telling Time to the Half Hour pp. 194–195

1. a) 9 o'clock   b) 6 o'clock

2.

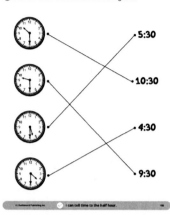

Exploring Digital Clocks p. 196

1.

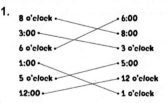

2. a) 12:00   b) 2:30

A.M. and P.M. p. 197

1. 7:00 A.M.   **2.** 3:00 P.M.

Months of the Year p. 198

1. January **2.** February **3.** March   **4.** April **5.** May **6.** June
7. July **8.** August **9.** September **10.** October   **11.** November
12. December

Reading a Calendar p. 199

1. November   **2.** Monday   **3.** Tuesday   **4.** Friday, November 19

Show What You Know! pp. 200–201

1.
8 o'clock — 6:00
3:00 — 8:00
6 o'clock — 3 o'clock
1:00 — 5:00
5 o'clock — 12:00
12:00 — 1 o'clock

2. a) 4 o'clock, 4:00
   b) 10 o'clock, 10:00

3. a) 10   b) 4   c) 12   **4.** a) 1:00   b) 2:00 P.M.